PUBLISHED BY

BOOKSURGE, LLC
5341 Dorchester Road, Suite 16
North Charleston, South Carolina 29418, USA

Ordering Information: www.Booksurge.com
Phone Number: 1-866-308-6235

First Edition

2003

A NOVEL THESIS

CHEMISTRY OF STABLE NEGATIVE CHARGE
AND
IT'S APPLICATIONS

WORK DONE BY THE AUTHOR IN PARTIAL FULFILLMENT
OF THE REQUIREMENTS FOR THE DEGREE OF
DOCTOR OF PHILOSOPHY FROM
1966-1969
AT UNIVERSITY OF WINDSOR
WINDSOR, ONTARIO, CANADA

BY

RAJENDAR K. SINGAL

PRESENT ADDRESS: 9084 ALPINE PEAKS AVENUE, LAS VEGAS, NV 89147

The thesis was originally submitted under the title, "Formation of Structures With The Negative Water Cluster". It was submitted by the author to the Faculty of Graduate Studies of the University of Windsor, Windsor, Ontario, Canada, in partial fulfillment of the requirements for the degree of Doctor of Philosophy. After finishing the required coursework in Physical, Analytical, and Polymer Chemistry, accomplishing mostly A grade in graduate level courses, thesis was submitted on December 9, 1969. The thesis was not accepted.

THE BOOK IS DEDICATED TO MY WIFE PROMILA AND OUR THREE CHILDREN, RAHUL, VINEY AND SANGEETA SINGAL

ABOUT THE AUTHOR

Rajendar K. Singal was born at Moga, Panjab, INDIA, on January 3, 1939. He graduated from Panjab University, Chandigarh, India, in M.Sc. (Honors) in Chemistry, in 1960. After working for Oil exploration industry for 5 years in India, he came in 1966 to University of Windsor, Windsor, Ontario, Canada, for Ph.D. degree in Chemistry. After finishing coursework accomplishing mostly A grade in graduate level courses, he submitted this thesis on December 9, 1969. He has worked for more than 30 years in USA in various Industries, such as Polymers, Wastewater Treatment and Chemical Analysis of Environmental Samples, Quality Assurance of Auto Parts, Rubbers, Lubes, Adhesives, Oils, Steels etc. More recently, he has worked on EPA projects on Quality Assurance of Environmental Samples Analysis Data, and Data Audits.

He has written this book for future scientists in order to advance the knowledge in the fields of Superconductivity at room temperature, Ambient Temperature High Energy Density Batteries and Solar Cells, Anionic Polymerization, based on the concept of Stable Negative Charge. The author firmly believes that major discoveries in science can be made in the future, by using the concept of Stable Negative Charge.

POTENTIAL APPLICATIONS OF STABLE NEGAATIVE CHARGE WHICH IS KNOWN AS HYDRATED ELECTRON

The references covered in this book are up to the year 1969 when the thesis was written. The author has published this book in order to draw the attention of scientists of the world to use the experimental data covered in this book for future research. The author envisions three main applications of the stability of Hydrated Electron. Research experiments can be designed on the basis of stable negative charge to investigate the possibility of SUPERCONDUCTIVITY at room temperature. It has been reported in several publications in literature that Superconductivity of several solid superconductors is achieved at low temperatures such as at liquid nitrogen temperatures or higher. It has yet to be proven in a single case about the existence of superconductivity at room temperature. The evidence of the stability of negative charge under certain conditions as expressed in this thesis can be utilized for further experimentation to find the Superconductivity phenomena at room temperature.

Second application is envisioned for the experimentation on High Energy Density Batteries for usages in cars, and other gadgets. Because in this thesis, there is evidence of the stability of negative charge, electrodes and electrolytes can be designed in several systems to provide high energy density batteries in deaerated electrolyte systems such as Lithium/Water, Zinc/Silver Oxide, Lithium/Conducting Polymers, Lithium/Sulfur Dioxide, Lithium/Mixture of Non-Aqueous and Aqueous Electrolyte. These are some examples of ambient temperature battery systems, which can be experimented based on the stability of hydrated electron.

Third application is envisioned for producing Polymers by the method of ANIONIC POLYMERZATION based on the stability of the negative charge and structure breaking aspects by the Monomers of stable hydrated electron.

THE UNIVERSITY OF WINDSOR

FORMATION OF STRUCTURES

WITH THE NEGATIVE WATER CLUSTER

A THESIS

SUBMITTED TO

THE FACULTY OF GRADUATE STUDIES

IN PARTIAL FULFILMENT

OF THE REQUIREMENTS

FOR THE DEGREE OF

DOCTOR OF PHILOSOPHY

DEPARTMENT OF CHEMISTRY

BY

R.K. SINGAL, M.Sc.

WINDSOR, ONTARIO

DECEMBER 9, 1969

Table of Contents

Chapter XI   Publications:
(1): Formation of Structures With the Negative Water Cluster, $n(H_2O)^{2-}$
            Part I – Photochemical Studies
Singal R. K., Indian J. Chemistry, Vol. 10, July 1972, pp. 718-723

(2): Radiolysis of KI – N_2O System;
Singal, R. K., Indian J. Chemistry, Vol. 9, No. 7, 1971, pp. 724-726

(3): Letter to the Editor, Chemical & Engineering News, March 1,1971,p.3
(4): Letter to the Editor, Chemical & Engineering News, May 8,1989,p.3

Figure Legend

Tables

Foreword

There has been an enormous growth of radiation chemistry over the past decade or so. The rates and mechanisms of the reactions in different systems are among the current interests in this field. There has been in the past and is going on presently an extensive investigation of the aqueous systems. The discovery of the novel species 'the hydrated electron' in 1962 increased the pace of the research and understanding of the aqueous systems at an alarming rate. The present research work described in the following few pages in this thesis suggests a rethinking of the study of the aqueous systems and 'the hydrated electron' and if the scientists agreed with the author partially if not completely, this thesis offers a breakthrough in understanding of aqueous radiation chemistry and from now onwards the research projects shall assume a new role in a completely different direction.

The author gratefully acknowledges the abilities and capabilities of his research adviser Dr. R. C. Rumfeldt who introduced him to the study of radiation chemistry. Due to the widely divergent views on the interpretation of the research data between the author and his research director, the former has made an independent effort in the interpretation of his research work. The author wishes that he were agreeable to the conventional rates and mechanisms based on the kinetic reactions of hydrated electrons which was the sole view of his research director. Looking at the experimental facts, there is no way out but to accept the challenge of convincing the scientific world that the concept of 'the hydrated electron' has failed in solving problems in aqueous radiation chemistry.

As shall be evident, the author's views presented in his thesis are based on the experimental facts which he obtained. Some of the radiation-chemical data appears to be not in agreement with the well established and accepted facts published in the literature. The author expresses confidence in the accuracy of the data which should be checked by others by doing in the same way as he did. It is not the humility that counts in scientific investigations but the ability, capability, pride and confidence of oneself which should be prerequisites for the Ph.D. degree. If the work expresses that previous investigations regarding the experimental data and or its interpretation are wrong, this should not be interpreted as the arrogance of the author to publish his work as he has obtained it. There are only two sides of the experimental work, either it is right or it is wrong and there is no midway.

The author appreciates the facilities for the radiation-chemical work provided by Atomic Energy Commission, Limited, Chalk River (Ontario) and is specially grateful to Dr. W. A. Seddon for his daily help in the lab which made the work possible.

The author expresses his indebtedness to the University of Windsor and her Chemistry Department for providing an opportunity to work and thus mix with the Canadian Society also.

Finally, he is thankful to his wife Promila for her patience and smile, which were constant sources of encouragement to him.

Chapter - I

Introduction

I - INTRODUCTION

<u>1.1</u>

Radiation Chemistry traces its origin to Roentgen's discoveries of 1895; however, it remains nameless until 1942. As a contemporary discipline it can be defined as that branch of chemistry which deals with the study of the chemical actions induced in matter by the absorption of high energy or 'ionising' radiations.

The absorption in matter of a high-energy particle or photon and any secondary particles that may be generated concomitantly is an exceedingly fast phenomenon, and is essentially completed before any response of the medium has time to develop. Energy loss by the particle(s) gives rise to the formation, along the particle track, of activated molecules which are chiefly in ionised or electronically excited states. These excited molecules are called primary products, and the period during which they are formed may be termed the physical stage of radiation action.

The primary products may be either extremely unstable or highly reactive and thus promptly undergo secondary reactions. This second, or physicochemical stage may comprise a single step, or a complex sequence of them.

Ultimately, the system attains thermal equilibrium, the primary products having been converted to stable moldcules (some of which may be reaction products, i.e., molecules different from those originally present) and to chemically reactive species (principally free atoms and radicalls). It thus enters a third, or chemical stage, in which these reactive species proceed to react with each other, or with the milieu.

In a biological system there follows a fourth, or biological stage which encompasses the sequential response of the organism to the chemical products of irradiation, through a hierarchy of organisational levels.

Although these four stages are by no means sharply demarcated, they provide an enlightening basis for analysing the complex succession of elementary events that are provoked by absorption of high energy radiation. It is characteristic that the duration of each stage is very small compared with that of the subsequent one. Actual durations depend very much upon the medium; order of magnitude values for a typical aqueous system are: physical stage, 10^{-13} second; physicochemical stage, 10^{-10} second; chemical stage, 10^{-6} second. The biological stage may, of course, extend over many years.

1.2

Primary Processes:

The primary processes of radiation chemistry are individual acts of energy transfer to atoms and molecules which lie in the vicinity of the path of a charged atomic particle. We are concerned here with interactions of C_o^{60} gamma rays with water. A brief description of the essential features of the primary energy loss processes is given as follows:

1.2.1

Photo-electric Effect:

In this process, all of the photon energy is transferred to a single bound electron (1, 2, 3). The absorption coefficient, T_a, varies approximately as $\lambda^3 z^4$ (where λ is the wave-length of the photon and Z ia the atomic number of the attenuating atom) and falls off rapidly with increasing photon energy. Thus, the photoelectric effect is most important for low energy photons and is usually the predominant process at photon energies comparable to the electron binding energies in atoms. It is an unimportant process for low Z materials when the principal mode of interaction with matter is by gamma rays in the energy range 0.2 to 4 MeV.

1.2.2

Compton Recoil Effect:

The principal mode of interaction with matter of gamma rays in the energy range (0.2 to 4 MeV) is by the compton process. In it, a photon interacts with an electron which may be loosely bound or free so that the electron is accelerated and the photon deflected with reduced energy. The energy and momentum of the incident photon are shared between the scattered photon and the recoil electron as given by the formula (4):

$$E_r = \frac{E_o}{1 + (E_o/m_o C^2)(1 - C_{os}\theta)} \qquad \dots \dots \dots (1)$$

Where E_o is the incident photon energy; E_r, the scattered photon energy; E_c, the recoil electron energy and $m_o C^2$ is the rest energy of the electron. This equation shows that when the angle θ is small, the photon is scattered with little reduction in energy and that the greater the deflection θ, the greater the energy loss from the photon. The energy of the recoil electron is equal to the difference between the energy of the incident and scattered photon:

$$E_c = E_o - E_r \qquad \dots \dots \dots \dots \dots \dots \dots (2)$$

and may have any value from zero to a maximum which can be calculated from equations (1) and (2) by putting $\theta = 0$ or 180° respectively. For example, for C_o^{60} gamma-rays of incident energy 1 MeV if $\theta = 0^o$, the photon will be deflected with no reduced energy as $E_o = E_r$ according to equation (1) and the recoil electron will have zero energy according to equation (2); however, if $\theta = 180^o$, the energy associated with recoil electron will be 0.32 MeV ($E_o/m_o C^2 = 0.51$ MeV).

The probability of the photon being scattered with a definite energy or direction and the probability of compton interaction as a whole are given by the relationship derived by Klein and Nishina (5). From this relationship it can be assumed that for high photon energies and intermediate Z materials, absorption by the compton process will be directly proportional to the electron density. This postulate is particularly significant for dose rate calculations in the present investigation where aqueous solutions have been irradiated with C_o^{60} gamma-rays.

1.3

Yield Nomenclature:

G-value: In radiation chemistry, the product yield is normally expressed as a G-value which is defined as the number of molecules changed for each 100 electron-volts of energy absorbed. Since we deal with high energy (0.2 to 4 MeV) C_o^{60} gamma-rays in these type of studies, to express the numbers in the form of G-value is a convenient way of representing them; for example, in calculating dose rates for dosimetry $G(Fe^{+3}) = 15.6$. Furthermore, the primary yields say of product 'X' are represented as G_x, whereas the measured ones as $G(X)$.

Φ-value: In photo-chemistry, the analogous yield is represented by the Φ-value which is defined as the number of molecules changed per quanta of light absorbed.

1.4

Radiolysis of Water:

A. O. Allen (6) has stated that the radiolysis of water must be recognized as one of the best understood topics in the radiation chemistry of condensed systems. This statement is perhaps applicable in respect to reaction mechanisms in aqueous systems, but the question of the origin and yields of the primary radiolytic species is yet to be satisfactorily resolved. It is appropriate therefore to present some of the essential features of the radiolysis of water.

The initial experiments with water revealed that with high Linear Energy Transfer (LET) (The rate of energy loss is generally expressed in terms of the linear energy transfer, or LET which is defined as 'the linear rate of loss of energy (locally absorbed) by an ionising particle traversing a material medium'.) irradiations such as by Alpha and Beta particles, the decomposition was evident (8), unless the water was irradiated in a vessel having a large evacuated volume over the water (9). The early work of Fricke initiated the development of the modern radiation chemistry of water. Fricke noted that whereas x-rays had little effect on pure water, considerable chemical reaction could ensue in the presence of dissolved solutes (10). He assumed that irradiation produced 'activated' water.

From studies made during the war years, it was concluded that irradiated water yielded both molecular and free radical products. The overall decomposition might thus be represented by the equation:

$$H_2O \longrightarrow H, \ OH, \ H_2, \ H_2O_2$$

The apparent stability of water to x-rays was found to be due to the attack of the free radicals on the molecular products to reform water by the chain reaction (6):

$$OH + H_2 \longrightarrow H + H_2O$$

$$OH + H_2O_2 \longrightarrow HO_2 + H_2O$$

$$H + H_2O_2 \longrightarrow OH + H_2O$$

The following reactions were also known to occur:

$$OH + H_2O_2 \longrightarrow HO_2 + H_2O$$

$$H + O_2 \longrightarrow HO_2$$

$$HO_2 + HO_2 \longrightarrow H_2O_2 + O_2$$

1.5

Molecular and Radical Yields:

The stoichiometry for the decomposition of water is given by the equation (6):

$$G(- H_2O) = G_H + 2G_{H_2}^M = G_{OH} + 2G_{H_2O_2}^M$$

Where the superscript 'M' denotes molecular yield, $G(-H_2O)$ represents total amount of water decomposed. These primary yields are based on the mechanisms postulated on the measured yields. The standard technique in determining the yields of each of the reaction products (H, H_2, OH and H_2O_2) employs the use of scavengers. For example, hydrogen yields can be obtained by intercepting the hydroxyl radicals before they can react with the molecular hydrogen. The iodide ion is an example of one such scavenger, although many others have been employed (11). Thus, the reaction

$$OH + I^- \longrightarrow I^0 + OH^-$$

protects the molecular hydrogen from the hydroxyl radicals. The hydrogen yield obtained in this way for Co^{60} gamma radiolysis is $G_{H_2}^M = 0.45_5$.

Another example of the use of a scavenger in an aqueous system is the determination of G_H in aerated 0.8 H_2SO_4 by ferrous ion oxidation (12). Each hydroxyl radical will oxidize one ferrous ion,

$$OH + Fe^{++} \longrightarrow OH^- + Fe^{+++}$$

and each peroxide molecule will remove two ferrous ions:

$$H_2O_2 + Fe^{++} \longrightarrow Fe^{+++} + OH^- + OH$$

$$OH + Fe^{++} \longrightarrow Fe^{+++} + OH^-$$

Hydrogen atoms react with dissolved oxygen to form the radical HO_2

which will subsequently cause the oxidation of three ferrous ions for each initial hydrogen atoms:

$$H + O_2 \longrightarrow HO_2$$

$$HO_2 + Fe^{++} \longrightarrow HO_2^- + Fe^{+++}$$

$$HO_2^- + H^+ \longrightarrow H_2O_2$$

The observed G-value for ferric ion formation is 15.6, thus the overall yield can be represented by:

$$G(Fe^{+++}) = 2G_{H_2}^M + 3G_H + G_{OH} = 15.6$$

From the stoichiometry of the reaction, $G(Fe^{+++})$ can be shown to equal $2G_{H_2}^M + 4G_H$ and by substituting $G_{H_2}^M = 0.45$, then $G_H = 3.65$. This particular system is the most commonly used chemical dosimeter and is known as the Fricke dosimeter. Several independent determinations (13) of $G(Fe^{+++})$ for this system have been made. Thus the value of $G(Fe^{+++}) = 15.6 \pm 0.2$ is accepted as a dosimetry standard for Co^{60} gamma-irradiations.

Samuel and Magee (14) and Platzman (12) proposed classical models to account for the experimental radical and molecular yields. These models were proposed about a decade earlier than the discovery of hydrated electron by Hart and Boag (15). The models do not predict at all the existence of hydrated electron. Samuel and Magee have assumed that the electron will ultimately be neutralized by the parent positive ion, after a random walk diffusion pattern. Platzman assumes that the fate of this electron will be to give H atom as follows:

$$\bar{e}_{aq} + H_2O \longrightarrow H + O\bar{H}_{aq}$$

The most serious criticism of the diffusion model is the necessity of employing numerous parameters of unknown values in order to achieve quantitative results. Thus, by a suitable adjustment of these values, it should almost always be possible to obtain a reasonable fit with the experimentally observed results.

More recently (16) Freeman proposed a non-homogeneous kinetic model which describes the kinetics of the scavenging of positive and negative species in spurs (a spur is a grouping of reactive intermediates that are close enough together that there is a significant probability that the intermediates in it will react with each other) during the radiolysis of liquids. He assumes a non-homogeneous distribution of radicals formed just after the deposition of energy. Those radicals which have escaped primary recombination and thus are far apart from their conjugate partners, will be readily scavenged at low concentrations of the solute. At high concentrations of the solute geminate recombination will also be scavenged, leading to higher yields of the same product. This model has been applied to irradiations in water with moderate success (16, 17).

Dainton and co-workers have reported numerous studies (18-23) of irradiated aqueous systems. On the basis of these studies, Dainton suggests (i) that the action of gamma-rays on water is inadequately represented by

$$G_{-H_2O} \, H_2O \longrightarrow G_e \bar{e}_{aq} + G_H H + G_{OH} OH + G_{H_2} H_2 + G_{H_2O_2} H_2O_2$$

in which the various G values are considered to be independent of solute concentration and (ii) that either there is also produced some other species which, in the absence of a sufficient concentration of an appropriate solute, rapidly reverts to water (e.g. H_2O^* or a radical pair) or, that at sufficiently large solute concentrations, the solute interferes with the recombination reactions (e.g. $\bar{e}_{aq} + OH \longrightarrow OH^-$) within the spurs. Their findings that increase in radical yield due to increase in pH or due to increase in scavenger concentration is in agreement with others (17, 24-29).

1.6

Electron Scavengers:

Electron capture by a neutral molecule may produce a stable negative molecule-ion intermediate. On the other hand, electron capture might proceed via a dissociative mechanism.

Broadly speaking, these two processes can be classified as:

$$AB + e \longrightarrow AB^- \qquad \text{(Resonance Capture)}$$

$$SF_6 + e \longrightarrow SF_6^- \qquad (30\text{-}33) \qquad (3)$$

$$I_2 + \bar{e}_{aq} \longrightarrow I_{2\,aq}^{-} \qquad (34) \qquad\qquad (4)$$

$$AB + e \longrightarrow A + B^{-} \qquad \text{(Dissociative Capture)}$$

$$N_2O + e \longrightarrow N_2 + O^{-} \qquad (35,\ 36) \qquad (5)$$

$$N_2O + \bar{e}_{aq} \longrightarrow N_2 + \bar{O}_{aq} \qquad (22) \qquad\qquad (6)$$

From mass spectrometric studies, the electron capture cross-section of $SF_6 \simeq 10^{-15}$ cm^2 (30-33), and that of $N_2O \simeq 10^{-19}$ (37). The peak of stable SF_6^{-} ion is also known from these studies (30 - 33). Its appearance potential is 0.08 eV and is used as an energy calibrant for monitoring the appearance potential of other ions. The high electron capture cross-section of the SF_6 molecule indicates that it should be an efficient electron capture agent as compared to nitrous oxide. The probability of electron capture by a neutral molecule in irradiated systems depends significantly on the threshold energy. Since the rate of energy loss of a subexcitation electron decreases with decreasing electron energy, molecules which possess high capture threshold energies may not be able to compete with electron energy moderation processes (12). Thus, those molecules which can capture electrons over lower energy ranges will have the highest probability of undergoing electron attachment reactions.

The extrapolation of all these arguments to condensed systems is rather dangerous and is still not well understood. After the discovery of hydrated electron (15), we know the left-hand side of these processes by studying them by fast reaction techniques; namely, flash photolysis and pulse radiolysis, and we assume the right-hand side from product analysis of steady-state radiolysis and photolysis systems. The intermediate formation is still an open question in these condensed systems and is known in only a few cases, such as:

$$\bar{e}_{aq} + I_2 \longrightarrow I_{2aq}^- \qquad (34) \qquad\qquad (7)$$

$$\bar{e}_{aq} + O_2 \longrightarrow O_{2aq}^- \qquad (38) \qquad\qquad (8)$$

1.7

Solvated Electron:

The present decade has seen an immense growth in radiation Chemistry with the discovery of the hydrated electron. In 1962, Hart and Boag,(15) found a broad band with λ max $\simeq 7200A^\circ$ by irradiating pure degassed water by the method of pulse radiolysis and interpreted this spectrum due to hydrated electron. At approximately the same time (39, 40), it was confirmed kinetically that the main reducing species was different from an 'H' atom in that it had unit negative charge. The main feature of the spectrum of the hydrated electron and trapped electron in the matrix is that in no case has any sign of any structure in the spectrum been detected (22). Though it is always a continuum, it has very large band width (about 25 K. Cal. for the hydrated electron) which is far more than can be accounted for by the effect of thermal motion on a single optical transition. Morever, the band is highly unsymmetrical on an energy scale, showing a surprisingly large tail on the high energy side. The extinction coefficient is always large, generally in the range 10,000 to 15,000 M^{-1} Cm^{-1} at λ max; and the oscillator strength is generally close to unity. The lack of structure, the asymmetry and lack of convergence argue against the origin of the band being np $\longleftarrow$ IS; the nature of optical transition responsible for the spectrum is still unknown. Alternative possibility is that the spectrum is really a charge transfer to solvent spectrum (41) and the observation of Anbar and Hart(42) that displacement in λmax for the electron band, caused either by change of medium or by addition of electrolytes, are exactly paralleled by shifts in the C.T.T.S. band of iodide ion affords some support for this hypothesis (A charge transfer to solvent, C.T.T.S. band of an ion is generally visualised as the shift in its absorption maxima due to the change in the environmental conditions of its medium which may be brought about due to pressure, temperature and ionic strength effects.)

Weiss (43) has pointed out that if the electron is still within the field of its parent ion, its spectrum should more closely resemble that of an F Centre and not that of a free hydrated electron. Experimental support for this view is claimed from the fact that the refined values of the λ_{max} for the transient species produced by flash photolysis of various ions, which are known to produce hydrated electrons, are not the same. Relevant values for the following ions are given in parentheses (22),

$$Cl^- \ (6,700), \ I^-(7,200), \ SO_3^{2-} \ (7,600), \ Fe(CN)_6^{4-} \ (8,000).$$

1.8

Methods for producing hydrated electron:

Electrons may be detached from molecules in three ways: Thermally, photochemically and radiation-chemically.

The thermal method requires the dissolution in water of molecules which spontaneously lose an electron to the water because the rate and energetics of the following reaction are favourable:

$$M \longrightarrow M^+_{aq} + \bar{e}_{aq}$$

The species coming into this category include the alkali and alkaline metals (44, 45).

When the above reaction is too slow to be achieved thermally, it is sometimes possible for an electron to be photodetached from a molecule M, according to:

$$M + h\upsilon \longrightarrow M^+_{solv.} + \bar{e}_{solv.}$$

By far the most convenient method is to irradiate the substance with ionising radiation, e.g. gamma-rays, beta particles, electron beams. In this thesis, we are concerned with electrons produced by Co^{60} gamma irradiation of water and that produced by photodetachment mainly in aqueous iodide solutions. Radiolytically, when a fast charged particle passes near a molecule, the coulombic interactions between the fast charged particle and an electron in a molecule ensure that, subject to certain quantum restrictions, energy is transferred from the particle to the electron. Molecules are thus electronically excited, and in some cases ionised. There are numerous advantages to this method, namely:

(1)   No competing solutes for the electron need be present other than H^+ and OH which are produced by the radiation in yields comparable to that of the electron.

(2)   The radiation is non-selectively abosrbed and therefore the presence of solutes chosen to react with the electrons will in no way affect the primary act.

(3)   Since the mechanism of the deposition of the energy is entirely independent of the translucency of the medium or of its state of aggregation, the method can be applied to coloured, cryst-alline or amorphous solids, as well as to liquids and, therefore, the technique of matrix isolation (for later study of the intermediates by either optical or magnetic spectroscopy) can easily be applied.

(4)   Possibly the greatest merit of this particular method is that it is now possible to give high dose pulses of very short duration, and therefore the technique of pulse radiolysis, but with even shorter pulse duration time can be applied to measure the absolute rate constants of reactions of the inter-mediate e^-_{aq} and OH.

In this thesis, the studies are concerned with hydrated electrons produced both by steady photolysis and radiolysis.

1.9

Previous studies on aqueous Iodide System:

1.9.1

Photo-chemical studies:

Dainton and Logan (46) investigated the photolysis of aqueous iodide ion using N_2O as the electron scavenger. They obtained $\Phi(N_2) = \Phi(I_3^-) = 0.23$, and thus assumed $\Phi(e_{aq}^-) = 0.23$. Jortner, Ottolenghi and Stein (47, 48) report this value as 0.29. Dainton and Sills (49) reported a value of 0.165 but this lower value was explained (46) perhaps as partly due to the effects of the accumulation of I_3^- ion in the unstirred solution and partly to the fact that the lamp used emitted more 3130A^o radiation than was allowed for in calculating the intensity. The above values are reported at 25°C' However, the quantum yield of iodine is temp. dependent (46, 48) and from these temp. dependent studies, the calculated activation energy 4.7 $\pm$ 0.5 K. cal./mole reported by Dainton and Logan (46) for the process $e_{aq}^- + N_2O \longrightarrow$ products is in good agreement with 4.9 K. cal./mole reported by Jortner, et al (48).

The maximum quantum yield for a non-chain photo-chemical process can be unity. However, it is generally agreed in the case of halides (Cl^-, Br^-, I^-) that the quantum yield of hydrated electrons produced on their photolysis is less than unity and the order is $Cl^- >$ $Br^- > I^-$ (46, 48, 50). The exact nature of the process to yield a quantum yield less than unity has not been explained but it is generally understood in terms of the competition processes between

the excited state of the anion to deactivate to the ground state and
the excited state of the anion to yield solvated electrons, namely:

$$A^-_{aq} \rightleftharpoons \overset{*}{A}{}^-_{aq} \longrightarrow A^o + \bar{e}_{aq}$$

Jortner, et al (48) explain the formation of a hydrated electron in the following manner: The investigations of the spectra of
I^-, $B\bar{r},Cl^-$ in aqueous solution (51) and the effects of changes in
temperature (52), Solvent and added ions (53, 54) are consistent with
the assumption that on the absorption of a light quantum the ground
state ion, occupying a cavity defined by the surrounding oriented
water molecules, goes over into an excited state in which, according
to the Frank-Condon principle, the atomic nuclei have preserved their
previous positions. This excited (C.T.T.S.) state consists of an
electron in a 2S orbital in the coulombic, spherically symmetrical,
field of the oriented solvent medium (55). The excited electron in
this orbital is confined over the first layer of water molecules of
hydration, its mean radius being $R_{ex} \cong$ 5-8A^o. They further state
that the results obtained (47, 48) indicate that there is a process
able to compete with the decay of the excited state. This process
leads to the formation of a halogen atom and a solvated electron,
$\bar{e}_{aq}$, in close proximity.

Smith and Symons (53, a) have developed a confined model for
the electron to explain the shifts in the absorption maxima of the
iodide ion due to environmental changes of the medium. They assume

a square well approximation for this model. Similar changes in absorption maxima in the case of Cl^-, Br^- have been explained in terms of this confined model (56). The confined model suggests that, to a first approximation, the excited electron is retained by the first layer of oriented solvent molecules, and the simple 'electron in a box' theory to describe this state is used. According to Smith and Symons (53, a), there are two reasons why it is thought that this model will result in a separation of the electron from iodine sufficiently great to result in the momentary formation of an iodine atom. The first is that the average distance of the solvent molecules in the first layer is, at room temperature, considerably greater than is normally postulated, and the second, that if an iodine atom is formed, there will be a marked shrinkage of the outer shell of electrons. The accepted values for the radii of iodide ions and iodine atoms are $2.16A^\circ$ and $1.3A^\circ$ respectively. This contraction will leave a large free space for the electron, and the presence of the iodine atom in the centre of a cavity now primarily concerned with the excited electron whould make very little difference to the energy of this electron. In general terms, the electron is thought to move in a discrete, centrosymmetric orbital defined largely by the potential field of those polarized solvent molecules which are oriented around the iodide ion. If the excited electron is confined to the solvent cavity, then it is possible that a simple square-well approximation might give some idea of the energy of this system. Ideally, it would be desirable to construct a well whose depth reflected the energy required to eject an electron totally from the solvent cavity. Since this energy is not known, the approximation that the walls are infinitely high is made. Provided the energy of the first level in the well is much less than its actual depth, this approximation should not make much difference. Pictorially, the retaining force may be thought of as the repulsion exerted on the excited electron either by the outer electrons of the solvent molecules, or by the outer sphere of negative charges on the solvent shell.

Dainton and Logan (46) state that the mode of destruction of excited aquated iodide ion is brought about by its collision with H_2O molecules, but that the dissociation reaction to yield hydrated electron may occur only if the configuration of the adjacent water molecules is suitable to allow the electron to escape from the potential well around the I atom and to reach a trap in the solvent where it may become hydrated. The dependence of the quantum yield of hydrated electrons ($Cl^- > Br^- > I^-$) produced from I^- by the absorption of $2537A^o$ light and from Cl^- and Br^- by $1849A^o$ light, on the type of halide has been explained by Jortner et al (48) in the following manner:

The transitions in the cases of I^- at $2537A^o$, Br^- and Cl^- at $1849A^o$ yield 2S state of the excited electron and the $2_{P_{3/2}}$ state of the halogen atom. The formation of e^-_{aq} from the primary excited state occurs in competition with the decay to the ground state which is rapid enough to prevent direct scavenging from the spectroscopic excited state. It involves a process of asymmetrization, in which the electronic charge is no more spherically symmetrical around the halogen atom (a

2S electron bound in the polarization field of the original ion), but a solvated, self-trapped electron in its lowest (1S) energy level. The efficiency of the asymmetrization of the first excited state to yield a solvated electron and a halogen atom increases in the series $I^-_{aq} > Br^-_{aq} > Cl^-_{aq}$. Two causes may contribute to this. The process of assymmetrization may involve the movement of the halogen atom which will be easier for the Cl atom than for the heavier Br or I atoms. A second cause may be a different rate of deactivation to the ground state, decreasing in the order I^-, Br^-, Cl^-.

1.9.2

Radiation-chemical studies:

Buxton and Dainton (18) studied the variation of radical and molecular yields in aqueous $KI-N_2O$ solutions over the pH range 0 to 14. As the pH is increased $G(N_2)$ at $[N_2O] = 1.5 \times 10^{-2}M$, increases from 0.25 at pH 0.1 to 3.2 at pH4, remaining constant to pH 10.8 when it again increases to 4 at pH 14. The data has been interpreted on the basis that $G_{e^-_{aq}} = 2.3$ and the excess $G(N_2)$ by increasing pH is due to the increase in radical yield and may be due to H_2O^* according to the following reaction:

$$H_2O^* + N_2O \longrightarrow N_2 + 2OH$$

This increase in $G(N_2)$ with high pH is in agreement with investigations of systems different from KI in the same laboratory (21, 57, 59) and with investigations in the other laboratories (25, 29).

Hayon (24) studied this system with nitrous oxide at natural pH and obtained results in fair agreement with Dainton and Buxton (18).

1.10

Purpose and scope of the present work:

Present studies were undertaken to:

a) Investigate the high $G(N_2)$, either by the increase of pH or by increasing scavenger (N_2O) concentration.

b) Introduce another radical scavenger, namely SF_6 and study the mechanisms of the reactions involved.

Sulphur-hexafluoride is known to be one of the most inert inorganic molecules. Its high electron affinity, however, has rendered it an extremely valuable specific electron scavenger in the radiolysis of gaseous (61), liquid hydrocarbons (62) and of water vapour (63). SF_6 has been previously used with N_2O and the suppression of the nitrogen yields produced by the reaction

$$\bar{e}_{aq} + N_2O \longrightarrow N_2 + O_{aq}^-$$

was explained as being due to electron scavenging by SF_6. However, no quantitative or mechanistic studies had been attempted with the exception very recently of the work of Asmus and Fendler (60).

It has been postulated by Dainton and coworkers that H_2O^* produced by radiolysis has a long life-time, perhaps being in the triplet State. On the other hand, only hydrated electrons and iodine atoms are produced by 2537A° photolysis of aqueous I^- by a well defined primary process. Thus, if a species other than hydrated electron is also produced in the gamma radiolysis due perhaps to a complicated primary process and its reaction kinetics differ significantly from hydrated electrons, as determined photochemically OR either N_2O or SF_6 is selective in scavenging this species, it could be possible to make a quantitative measure employing a suitable kinetic analysis.

Another obvious extension of these studies was to investigate the behaviour of other radicals towards SF_6 in order to gain a general pattern of its radical scavenging characteristics. Only photolysis of Cl^-, Br^-, CN^-, SCN^- and $Fe(CN)_6^{4-}$ has been attempted due to lack of facilities for radiation-chemical work available here.

Chapter - II

Experimental

II - EXPERIMENTAL

2.1

Materials:

Solutions were prepared from quandruply distilled water. Triply distilled water was kept under continuous reflux over alkaline permanganate and collected just prior to the preparation of solutions.

Nitrous oxide and sulfurhexafluoride (Matheson) were purified by passage through a column of potassium hydroxide pellets to remove traces of CO_2 and followed by trap-to-trap distillation, several cycles of freezing at $-196°C$, evacuating and revaporising. One such cycle was completed just before each experiment.

Analar grade reagents (Fisher) were used without further purification for preparing the solutions.

2.2

Apparatus and Procedure:

The reaction vessel used for photochemical studies is illustrated in figure 1. During irradiations it was immersed in a tank thermostated to $25\pm1°C$ with a horizontal quartz window, below which was placed the quartz cell, A. The solution was agitated by means of the magnetically operated stirring paddle, B, to minimize the accumulation of products in the regions where light absorption is highest.

The vessel was cleaned with permanganic acid followed, after rinsing, by a mixture of nitric acid and hydrogen peroxide. After

Figure 1

Reaction vessel used for photochemical studies, quartz cell, A, stirring paddle, B, bulb, c, standard tapper joint, D, side arm, F.

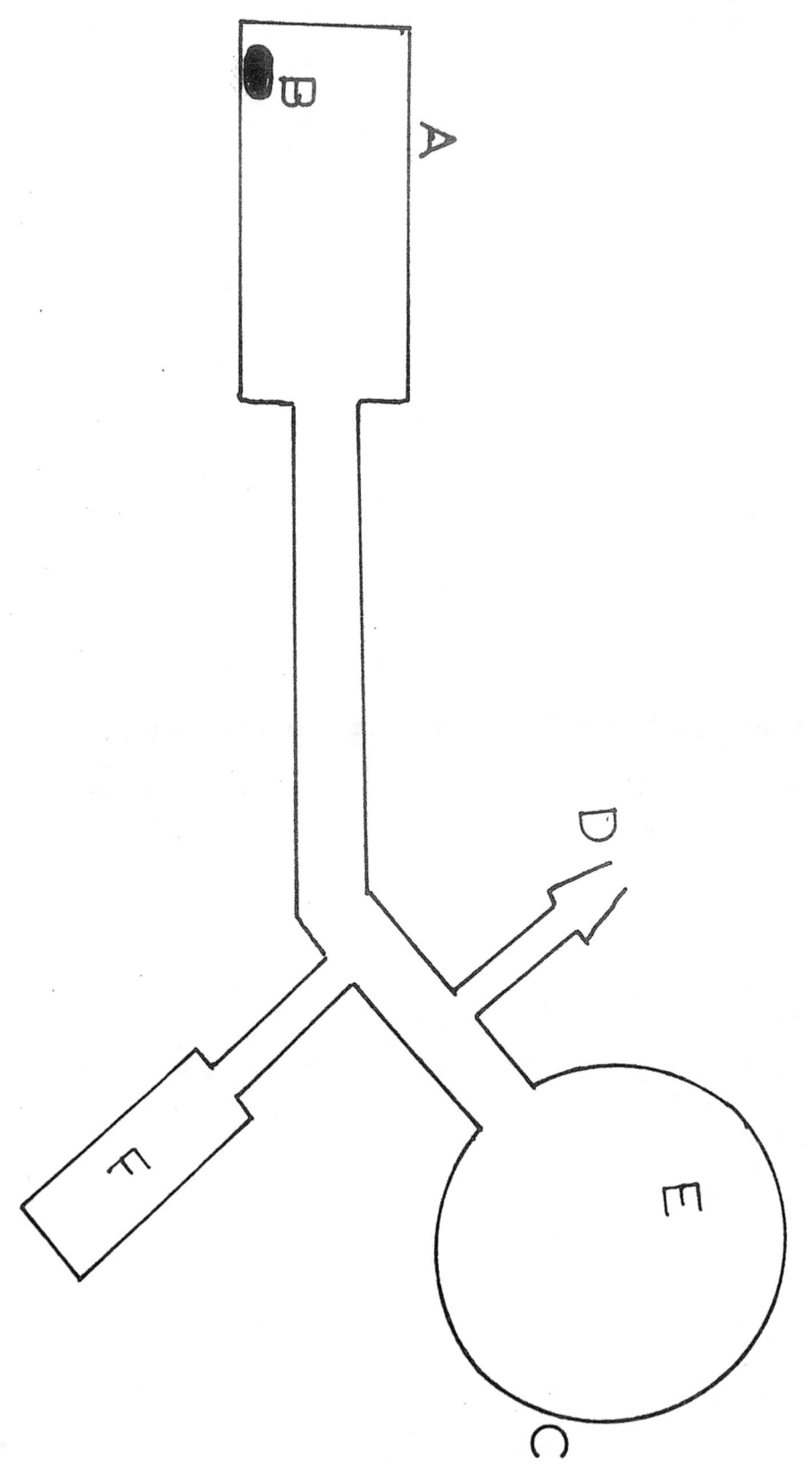

some hours contact with the later liquid, the cell was rinsed ten times with triply distilled water and then with quadruply distilled water and finally dried in an oven at $110^{\circ}C$.

A 35 c.c. sample of $5\times10^{-3}M$ solution to be photolysed was poured into the 100 ml. bulb, C, a joint and a right angle tap fitted to the inlet, D, and the solution deaerated by four or five cycles of freezing, evacuating and thawing. A trap immersed in a bath of solid carbon dioxide and acetone served to mimimize the transfer of mercury vapour to the reaction vessel from the space above the mercury manometer, when the nitrous oxide was added from the storage bulb. SF_6 was added in the same manner, except a bath of salt and ice was used in place of a bath of solid carbon dioxide and acetone, to minimize the transfer of mercury vapour. For studies with N_2O or SF_6, the solution was equilibrated by constant shaking of the cell and the equilibrium partial pressure was noted after about 10 to 15 minutes.

For competition studies with nitrous oxide and SF_6, the desired amount of SF_6 was measured in the usual way, however, it was then distilled into an empty bulb and maintained at liquid nitrogen temperature. The distillation cycle was repeated several times to ensure complete transfer of the SF_6. The vacuum line was pumped after this process to make it free from any gases. The nitrous oxide was equilibrated at its desired pressure. The gas and the solution in the reaction vessel was frozen with liquid nitrogen; then the previously measured SF_6 was distilled back into the reaction vessel. The vessel was then partially immersed in hot water and when the solution was at approximately room temperature, it was put into the thermostated bath and the system was allowed to stand for at least 15 minutes before irradiation.

2.3

Actinometry:

The radiation source used was a model PCQ 0115 (Ultra Violet Produce Inc., San Gabriel, California), low pressure mercury grid lamp, emitting mainly the 2537A$^{\circ}$ line and only a small amount of the 1849A$^{\circ}$ line with perhaps some longer wavelength lines. The source was used in conjunction with a filter of triply distilled water, 2 cm. deep, to remove low wavelength radiation, such as the 1849A$^{\circ}$ mercury line. Light intensities were measured for reaction vessel given in figure 1 using a solution of the ferrioxalate actinometer (64) in the same cell placed in the same position. Pure potassium ferrioxalate was prepared by mixing 3 volumes of 1.5M A.R. potassium oxalate with 1 volume 1.5M A.R. ferric chloride with vigorous stirring. The pre-cipitated potassium ferrioxalate was recrystalised three times from warm triply distilled water and the crystals filtered and dried in a current of air at about 50°C. The composition is reported to corre-spond to $K_3Fe(C_2O_4)_3 \cdot 3H_2O$. The 6 x 10^{-3}M actinometric solution was irradiated using the radiation source in conjunction with a filter of triply distilled water. The concentration of Fe^{2+} was calculated from its optical density at 5100A$^{\circ}$ using the extinction coefficient 1.11 x 10^{4}1M^{-1}Cm^{-1}. The extinction coefficient was determined from the cal-ibration graph prepared for Fe^{2+} by the method of Hatchard and Parker (64). The calculated dose rate for 2537A$^{\circ}$ light was 1.08 x 10^{20} quanta/min. using $\phi(Fe^{2+})$ = 1.25 for this line. Corrections were made for the effect of the longer wavelength lines absorbed by the actinometric solution by replacing the filter cell between the lamp and the window by a similar one containing a thoroughly degassed 5×10^{-3} M solution of KI, which selectively removes lines of interest. It was found that excluding those wavelengths absorbed by the water filter 76.5 % of the actinic effect of the lamp was due to $2537 A^\circ$ light.

2.4

Light fraction correction:

Actinometry for the $1849 A^\circ$ line using the reaction vessel shown in Fig. 1 was done by the system suggested by Dainton and Fowles (65). Aqueous solutions of N_2O at 2.08×10^{-2} M N_2O were photolysed in the reaction vessel. The yield of nitrogen as a function of dose was linear with a slope of 2.16×10^{-6} M/min. Assuming a quantum yield of unity for nitrogen formed by the direct photolytic decomposition of N_2O, a dose rate of 3.1×10^{18} quanta/min. for $1849 A^\circ$ light was calculated after correcting for the light fraction absorbed by N_2O by the following relationship:

$$\Phi(N_2) = \Phi^\circ(N_2) \cdot \frac{\epsilon_{N_2O} \; [N_2O]}{\epsilon_{N_2O} \; [N_2O] + \Sigma \; H_2O}$$

where $\Phi(N_2)$ is the observed quantum yield of nitrogen, $\Phi^\circ(N_2)$ is the primary quantum yield of nitrogen (unity in this case); ϵ_{N_2O} is the extinction coefficient of N_2O, $\epsilon_{N_2O} = 60$ l $M^{-1}Cm^{-1}$ (65) and $\Sigma H_2O = 1.8$ Cm^{-1} (65).

Thus, for $N_2O = 2.08 \times 10^{-2}M$,

$$\Phi(N_2) = 0.409$$

and the light intensity is:

$$\frac{2.16 \times 10^{-6} \times 6.02 \times 10^{23}}{0.409} = 3.1 \times 10^{18} \text{ quanta/min.}$$

2.5

High pressure studies:

Since the solubility of SF_6 is about 100 times less than N_2O (66, 67), an equilibrium partial pressure of 1 atm. of SF_6 only results in a concentration of $2.4 \times 10^{-4}M$. Thus, a high pressure system was designed for photochemical experiments (Fig. 2). A small cup, A, made of heavy wall pyrex was jacketted in a brass container and the top of the cup was sealed by placing the quartz window, B, between two O-rings on this cup, with a threaded metal disc, C, fitting the brass jacket, D. The pyrex side arm, E, of this cup was fitted with a teflon stop-cock which could stand pressure up to 15 atm. The gases were condensed into the reaction vessel from the gas bulb of known PVT. The total volume of the reaction vessel with the teflon stop-cock closed was 26.6 c.c. and that of the gas bulb was 497.6 ml. The concentration of the gaseous additives in the liquid phase were cal-culated from the known solubility data (66, 67) using:

Figure 2

Reaction vessel used for high pressure photo - chemical studies.
Pyrex cup, A, quartz window, B, metal disc, C, brass jacket, D, pyrex
side arm, E, stirring paddle, F.

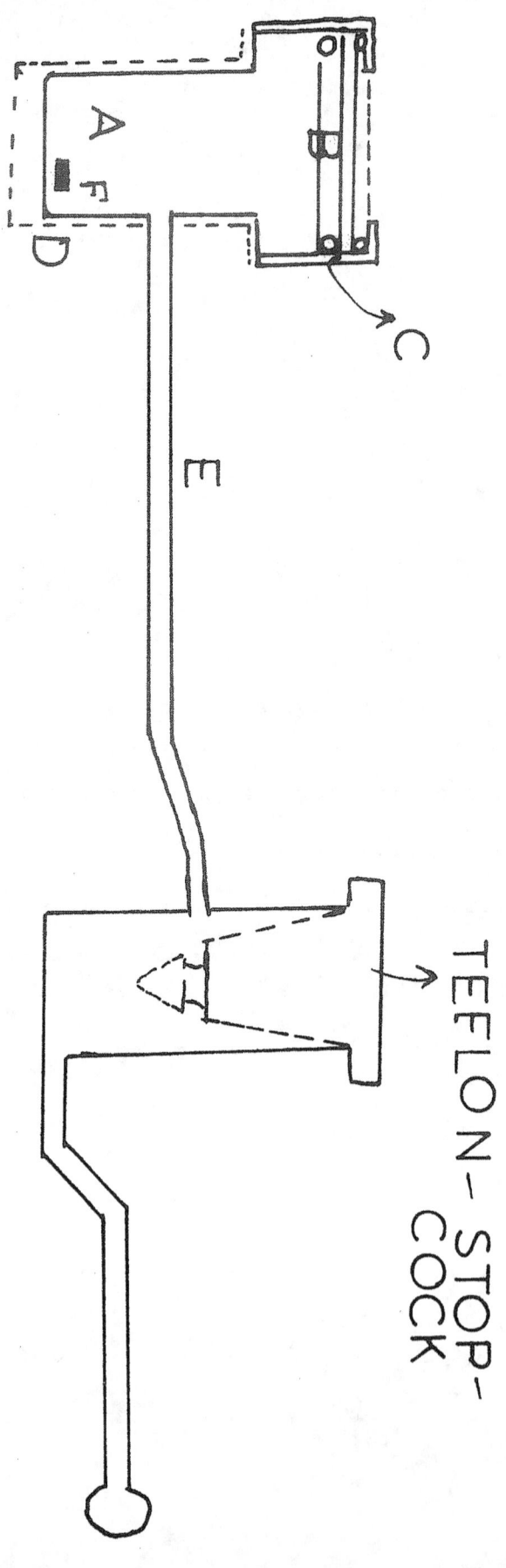

TEFLON-STOP-COCK

$$C_{gas} = \frac{\alpha \cdot .1000}{22400.760} \cdot \frac{P_o V_o}{(V_{gas} + V_{liq.})} [\underline{M}]$$

Where C_{gas} is the concentration of the dissolved gas in the solution after equilibration, α is the solubility of the gas in Cm^3 per Cm^3 H_2O, $P_o V_o$ is the amount of added gas in Cm^3 Torr, V_{gas} is the volume of the gas phase in Cm^3 and $V_{liq.}$ is the volume of the liquid phase in Cm^3. The cell was leak-tested before photolysis and was thermally equilibrated at room temperature by stirring the solution for half an hour with the stirring paddle, F. A special stand was designed to fit the cell. Studies were carried out using 2537A^o light with the iodide and ferrocyanide systems and with 1849A^o light with Cl^-, Br^- and SCN^- systems. The intensity of 2537A^o light for this reaction vessel was 8×10^{19} quanta/min. and that of 1849A^o light was 3.55×10^{18} quanta/min.

2.6

Example for using the light fraction correction:

When 1849A^o light is the photolysing line, the quantum yield of products must be corrected since only a fraction of the light absorbed is by the ion being photolysed with the remainder being absorbed by the water. The following extinction coefficients for the ions photolysed with 1849A^o light were used for these corrections:

$$\text{Chloride} = 3.8 \pm 0.3 \times 10^3 \text{ l } M^{-1}Cm^{-1} \tag{50}$$

$$\text{Bromide} \quad = \quad 1.3 \pm 0.1 \times 10^4 \, 1 \, M^{-1} Cm^{-1} \tag{50}$$

$$\text{Cyanide} \quad = \quad 6.7 \times 10^2 \pm 201 \, M^{-1} Cm^{-1} \tag{72}$$

$$\text{Thiocyanate} = 1.63 \times 10^4 \pm 1001 \, M^{-1} Cm^{-1} \tag{72}$$

Thus, for example, if the yield of fluoride was $0.66 \times 10^{-5} M$ by photolysing Cl^- solution and since light intensity is 3.1×10^{18} quanta/min.

$$\Phi(F^-) \quad = \quad \frac{0.66 \times 10^{-5} \times 6.02 \times 10^{23}}{3.1 \times 10^{18}}$$

$$= \quad 1.25$$

But $\qquad \Phi(F^-) \quad = \quad \Phi^\circ(F^-) \times \dfrac{\epsilon_{Cl^-} [Cl^-]}{\epsilon_{Cl^-} [Cl^-] + \Sigma H_2 O}$

Substituting the appropriate values

$$1.25 \quad = \quad \Phi^\circ(F^-) \cdot \frac{3.8 \times 10^3 \times 5 \times 10^{-3}}{3.8 \times 10^3 \times 5 \times 10^{-3} + 1.8}$$

Then $\qquad \Phi^\circ(F^-) \quad = \dfrac{1.25}{0.914} \quad = \quad 1.37$

In the following sections all quantum yields of fluoride obtained by photolysis with $1849 A^\circ$ light have been corrected using light fraction relationship.

2.7

Gamma radiolysis:

The reaction cell used is shown in figure 3. Cells, pipettes and storage flasks were cleaned by rinsing them successively with permanganic acid, distilled water, hydrogen peroxide containing nitric acid and finally, four times with triply distilled water. 10 ml. of 5×10^{-3} M KI solution was pipetted into the clean cell and deaerated by the procedures described earlier. Desired pressures of SF_6 and/or N_2O were obtained within the cell by the procedures described earlier and the concentration calculated from known solubility data of each gas (66, 67). The system was thermally equilibrated at room temperature (approximately $25 \pm 2^{\circ}C$) before irradiation.

2.8

Dose rate measurement:

Dose rates were measured by the Fricke dosimeter, using $G(Fe^{3+})$ = 15.6 ions per 100 eV for Co^{60} gamma rays. A dose rate of 7.6×10^{20} ev/l/min. was calculated from the quantity of ferric ion formed as measured by a direct reading of the optical density of the solution with an U.V. spectrophotometer at the wavelength $3050A^{\circ}$ and an extinction coefficient of 2201 at $25^{\circ}C$ for ferric ion at this wavelength was used in the calculations.

2.9

ANALYSIS:

Figure 3

Reaction cell used for radiation - chemical studies.

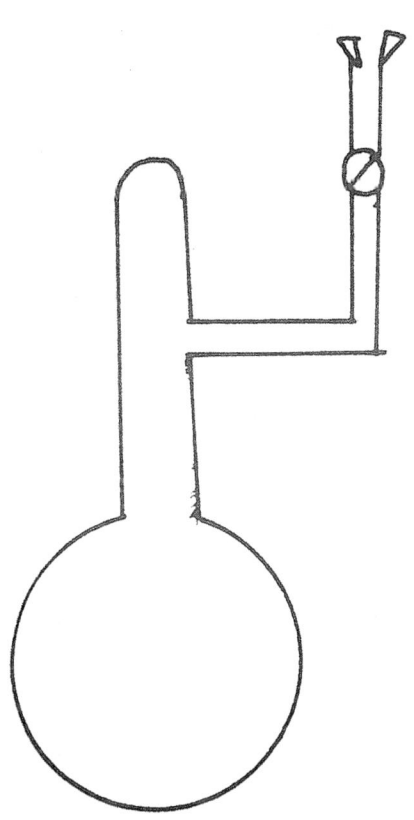

2.9.1

Gaseous products:

a) Liquid nitrogen as refrigerant of Water:

   After photolysis, the solution was frozen with liquid nitrogen and the gaseous products taken from the reaction vessel in three cycles of freezing, pumping and thawing, were passed through traps at -196°C, to remove condensable gases, and collected in a calibrated capillary tube by means of a mercury diffusion pump with a small dead space and a toepler pump. In all the experiments, the waiting period after freezing and during pumping the gas in the capillary tube was at least five minutes each time.

   In radiation-chemical studies, a different kind of apparatus (Fig. 4) was used for collecting the gases from the reaction vessel. After radiolysis, the solution was frozen with liquid nitrogen and the gaseous products taken from the reaction vessel in three cycles of freezing, pumping and thawing were expanded in the big bulb of the toepler pump. The volume of the total dead space as compared to the bulb was kept small so that the gases could be collected efficiently in the bulb of the toepler pump. Then the mercury was raised in order to push this gas in the capillary tube. This process was repeated several times. The gases, before going in the bulb of the toepler pump, were passed through a U-tube trap maintained at liquid N_2 temperature attached with a stop-cock after it (Fig. 4). This stop-cock was kept closed for about two minutes before the expansion of the gases in the bulb of the toepler pump, in order to condense completely the condensable gases in the U-tube trap.

Figure 4

Apparatus for the measurement of gaseous products in radiation - chemical studies.

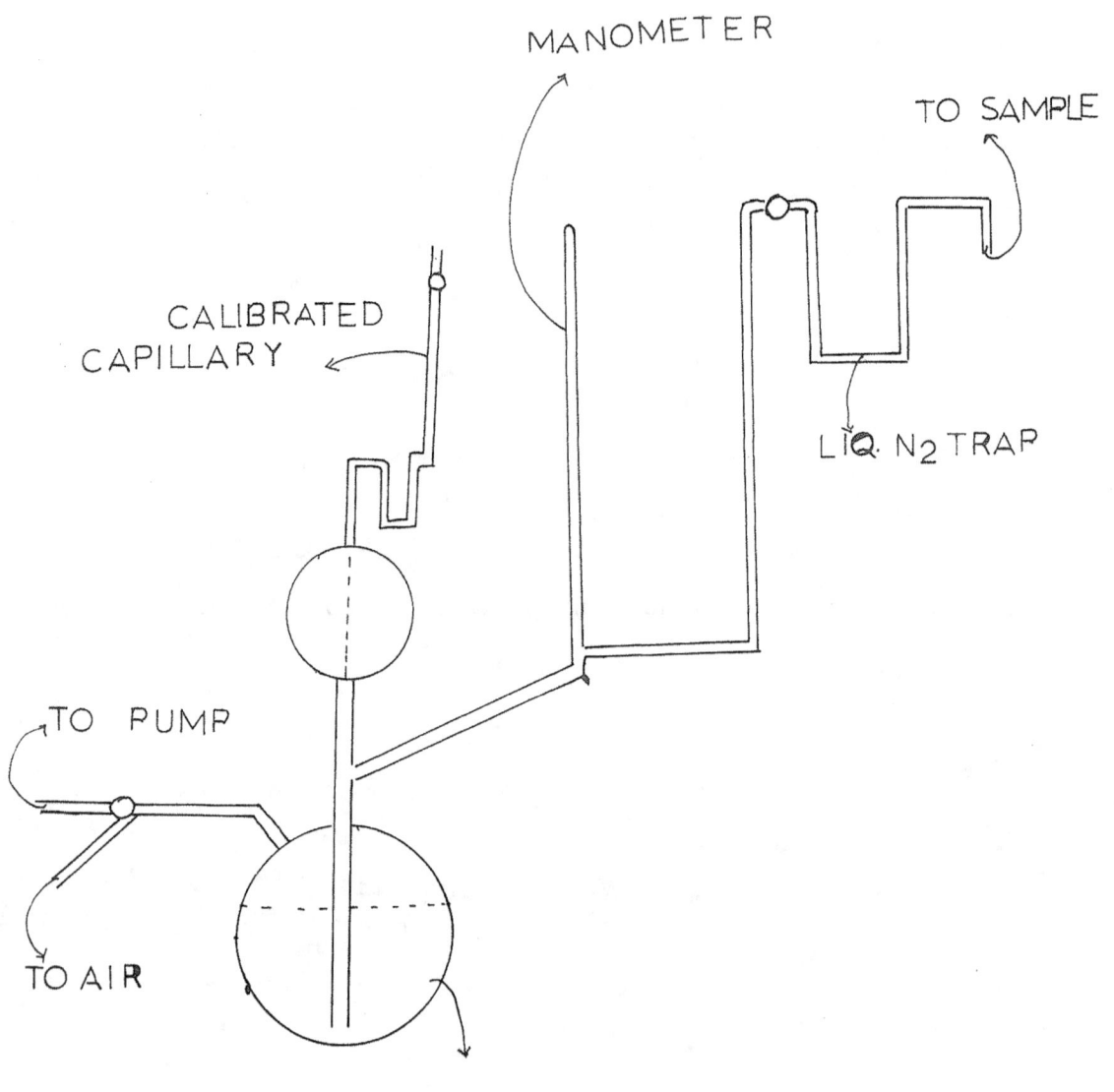

MANOMETER

TO SAMPLE

CALIBRATED
CAPILLARY

LIQ. N₂ TRAP

TO PUMP

TO AIR

MERCURY

b) <u>Dry Ice + Acetone as refrigerant of water</u>:

The process (a) was repeated using dry ice + acetone instead of liquid nitrogen as refrigerant of water.

c) <u>No refrigerant of water</u>:

In some experiments, no freeze-pump-thaw cycles were carried out and the gases were extracted by pumping at room temperature and passing them through two liquid nitrogen traps to freeze out the condensable gases and water vapour.

In photo-chemical experiments, quantitative analysis of the gaseous products was done by gas chromatography using a Varian Model 1521-1B Gas Chromatograph with a molecular sieve 5A column and a thermal conductivity detector. The instrument was calibrated for analysis of O_2 and N_2 using pure samples of each gas. N_2 was the only gaseous product in all experiments and the upper limit of reliable detection by the instrument was 0.5×10^{-7} moles, but the overall accuracy under the experimental conditions employed was not better than $\pm 0.5 \times 10^{-7}$ moles. In radiation-chemical experiments, the only gaseous products were nitrogen, hydrogen and oxygen which were analysied by mass-spectrometer and the accuracy achieved was within $\pm 2 - 3 \%$.

Blank experiments were carried out by preparing the sample in the normal way after which, without photolysis or radiolysis, the gases were measured in the capillary tube and analysed by gas chromatographic or mass-spectrometric techniques respectively. In the photochemical systems where relatively large volumes of solutions were used

and complete deaeration was difficult to achieve, residual N_2 was detec-

ted in the blanks. The average blank was $.5 \times 10^{-7}$ moles. This blank

correction was applied in all cases in photochemical experiments.

However, in the radiation-chemical experiments, the blank correction

was very small and the average correction was only 1×10^{-9} moles of

N_2 and oxygen was below the limits of detection of the mass-spectrometer.

2.9.2

Solution products:

a) Degassed yields:

Solution products were analyzed after the measurement of gaseous

products.

b) Non-degassed yields:

Solution products were also analyzed just after photolysis or radiol-

ysis without the measurement of gaseous products.

In photolysis, the only solution products were iodine, fluoride

and sulphate. In radiolysis, the only solution products were iodine,

fluoride, sulphate and hydrogen peroxide.

2.9.2.1

Iodine:

The iodine produced in the experiments exists in equilibrium with

I_3^- and the latter was measured spectrophotometrically at 3530A$^\circ$. In

some photochemical experiments, it was measured by putting the reaction

cell, with the stirrer transferred to position E, directly into the

sample compartment using the side arm F with the reaction vessel (Fig.1) and the same cell holder as for others. The total yield of iodine was calculated, using values of 2.64×10^4 $M^{-1}Cm^{-1}$ for the decadic molar extinction coefficient of the I_3^- ion (68) and 742 M^{-1} for the equilibrium constant $[I_3^-] / [I^-] [I_2](46)$.

2.9.2.2

Fluoride:

Fluoride yields were measured by an Orion Research Specific Ion Fluoride Electrode and the Orion Research Specific Ion Meter model 401. A calomel electrode was used as the reference in conjunction with the fluoride electrode. In some photo-chemical experiments, the fluoride yield was measured on the Orion Research Specific Ion Digital Meter model 801. The measurements of the fluoride yield by both the instruments agreed within $\pm$ 1%. A calibration curve (potential, MV vs concentration of standard fluoride solution supplied by Orion) was drawn up to 1×10^{-6}M and the product fluoride yield was computed from the standard curve. In order to keep the ionic strength in the standard fluoride solutions approximately the same as in the samples, they were diluted with a 5×10^{-3}M KI solution. The fluoride ion electrode exhibited a 59.0 MV change in potential for each ten-fold change in concentration over the range 10^{-1} to 10^{-5}M and was reproducible to $\pm$ 0.5 MV. For the range 10^{-5} to 10^{-6}M fluoride solution, the change was only 48 MV and this is attributed to the lower response of the electrode at this concentration of fluoride ion. The electrodes were calibrated daily with standard fluoride solutions which had approximately the same ionic strength as the irradiated samples. These calibrations gave excellent straight lines. The limit of detection by this method (69) is 10^{-6}M fluoride ion, and by using the microsample dish, determinations could be carried out on as little as 1.0 ml. of sample. In the pH region of 4.5-7, the electrode is completely selective for fluoride ion, however, at lower pH values, hydrogen ions form HF and HF_2^- which are not detectable by the electrode (70). In all these experiments, in no case did the pH of any solution go below 4.5.

2.9.2.3

Sulphate:

Quantitative analysis of the sulphate yields was not achieved due to irreproducibility of the data. For the analysis, some experiments were tried with the nephlometric technique (III) but no meaningful result could be obtained. However, sulphate was detected by a solution of barium chloride.

2.9.2.4

Hydrogen peroxide:

H_2O_2 was determined in some radiation-chemical experiments by the method of Allen, Hochanadel, Ghormley and Davis (71).

2.9.2.5

Ferricyanide:

The concentration of $Fe(CN)_6^{3-}$ was determined by its absorption at 4200A$^{\circ}$ using the extinction coefficient 1030 ± 40 1 $M^{-1}Cm^{-1}$ at this wavelength (73). If monoaquo complex $Fe(CN)_5 \cdot OH_2^{3-}$ is also formed in the photolysis of $Fe(CN)_6^{4-}$ systems, it could not be measured

as its spectra overlaps that of $Fe(CN)_6^{3-}$ (73).

Chapter - III

Results

Photo - chemical

III - RESULTS

All photo-chemical experiments have been done at the natural pH of the deaerated solutions.

3.1

KI :

5×10^{-3} M KI solution was irradiated with 2537A° light for several minutes. No products were formed and this observation is in agreement with Dainton and Logan (46).

3.2

KI - N_2O :

A series of experiments were carried out using 5×10^{-3} M solutions of KI containing different $[N_2O]$. The quantum yield of iodine (non-degassed, measured in the side arm without opening the reaction vessel) and nitrogen (technique 2.9.1 C, i.e. no refrigerant of water) were equal and the results (fig. 5) at 1.5×10^{-2} M N_2O concentration are in agreement with Dainton and Logan (46). The results (fig. 5) at N_2O higher than 2.6×10^{-2} M have not been previously reported for this type of photo-chemical studies.

3.2.1

The quantum yield of nitrogen (technique 2.9.1 b) at 1.5×10^{-2} M N_2O was 0.285 ± 0.005 and that of iodine (measured after the nitrogen analysis) was 0.265 ± 0.005. These results are in agreement with Jortner et al (47, 48).

Figure 5

$\Phi(\bar{e}_{aq})$ as a function of [scavenger]. $\triangle$ $\Phi\frac{(F^-)}{6}$ = $\Phi(\bar{e}_{aq})$,

$\bullet$, $\Phi(I_2)$ = $\Phi(N_2)$ = $\Phi(\bar{e}_{aq})$, $\square$, $\Phi(I_2)$ = $\Phi(N_2)$ = $\Phi(\bar{e}_{aq})$ from the work of

Dainton and Logan (46). SF_6 and N_2O are the scavengers used for the

present studies.

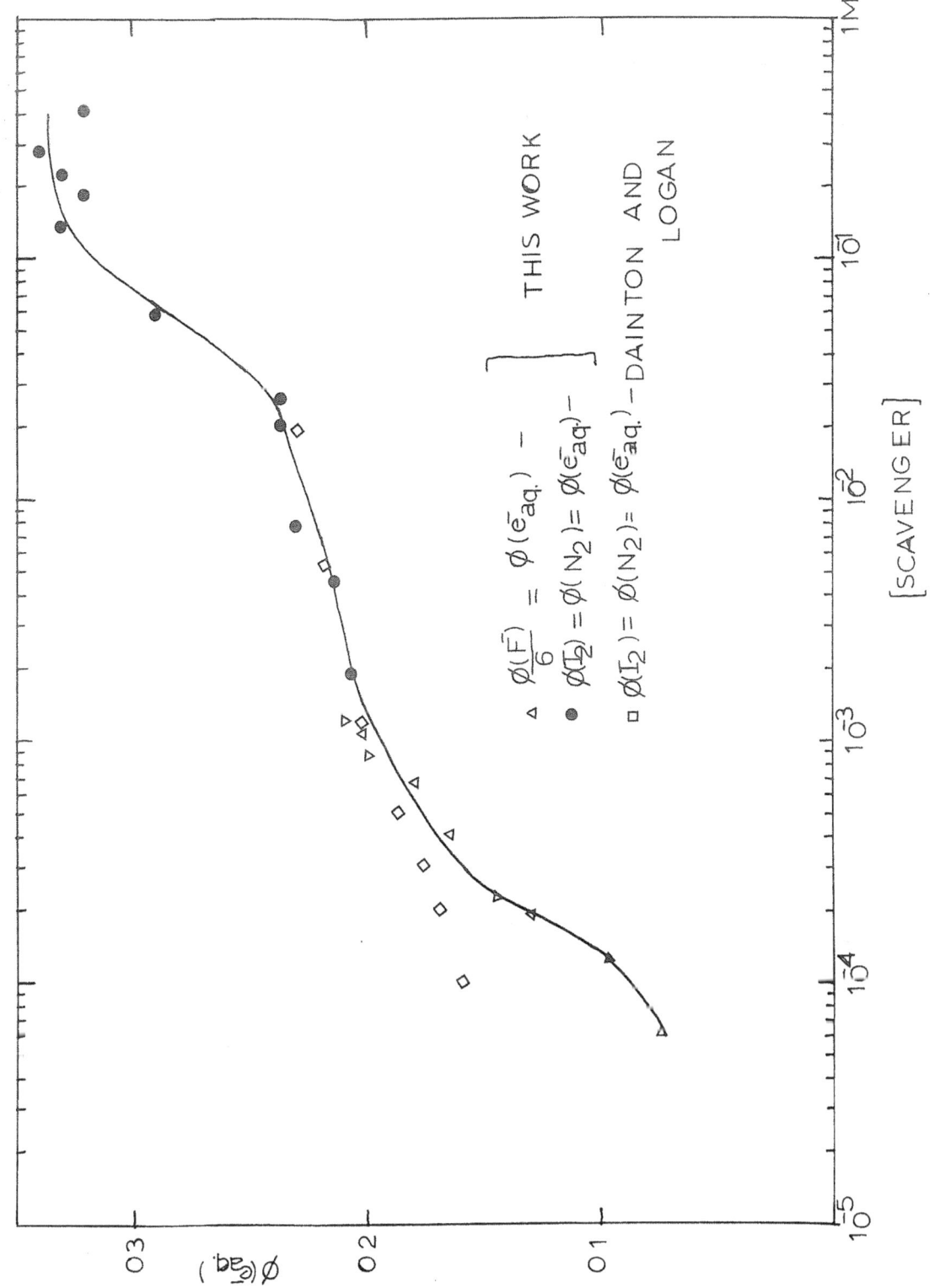

3.2.2

Some experiments were carried out using 5×10^{-3} M solution of KI containing 1.5×10^{-2} M N_2O and measuring N_2 using technique 2.9.1 a, and iodine after the nitrogen analysis. The yields of nitrogen were irreproducible and much lower than measured by techniques b and c. In the first freeze-pump cycle, (technique a) there was no nitrogen. In the second and third cycle, there was very little nitrogen. The behaviour of the first cycle was always reproducible but not in the second and third cycles. The quantum yield of iodine measured after the nitrogen analysis was 0.24 ± 0.01.

3.3

KI - SF$_6$:

The fluoride yield as a function of dose is shown in fig. 6 for $[SF_6] = 2.4 \times 10^{-4}$ M. The yield of fluoride was linear up to 1 minute of irradiation time and $\phi(F^-) = 0.87 \pm 0.03$ was calculated from the initial linear portion. The variation of the quantum yield of fluoride with $[SF_6]$ (fig. 7) was plotted and the maximum quantum yield of hydrated electrons, assuming $\dfrac{\phi(F^-)}{6} = \phi(e^-_{aq})$ achieved was only 0.22 at approximately 1.2×10^{-3} M SF$_6$ after which it appeared to drop. This feature of the decrease of quantum yield after a particular concentration of scavenger was not observed with N_2O (fig. 5). The quantum yield of hydrated electrons up to approximately 4×10^{-4} M SF$_6$ is lower than observed at the same concentrations of N_2O by Dainton and Logan (46) (fig. 5).

Figure 6

Yield of F^- as a function of irradiation time. $[SF_6] = 2.4 \times 10^{-4}$M, dose rate $= 1.08 \times 10^{20}$ quanta/min.

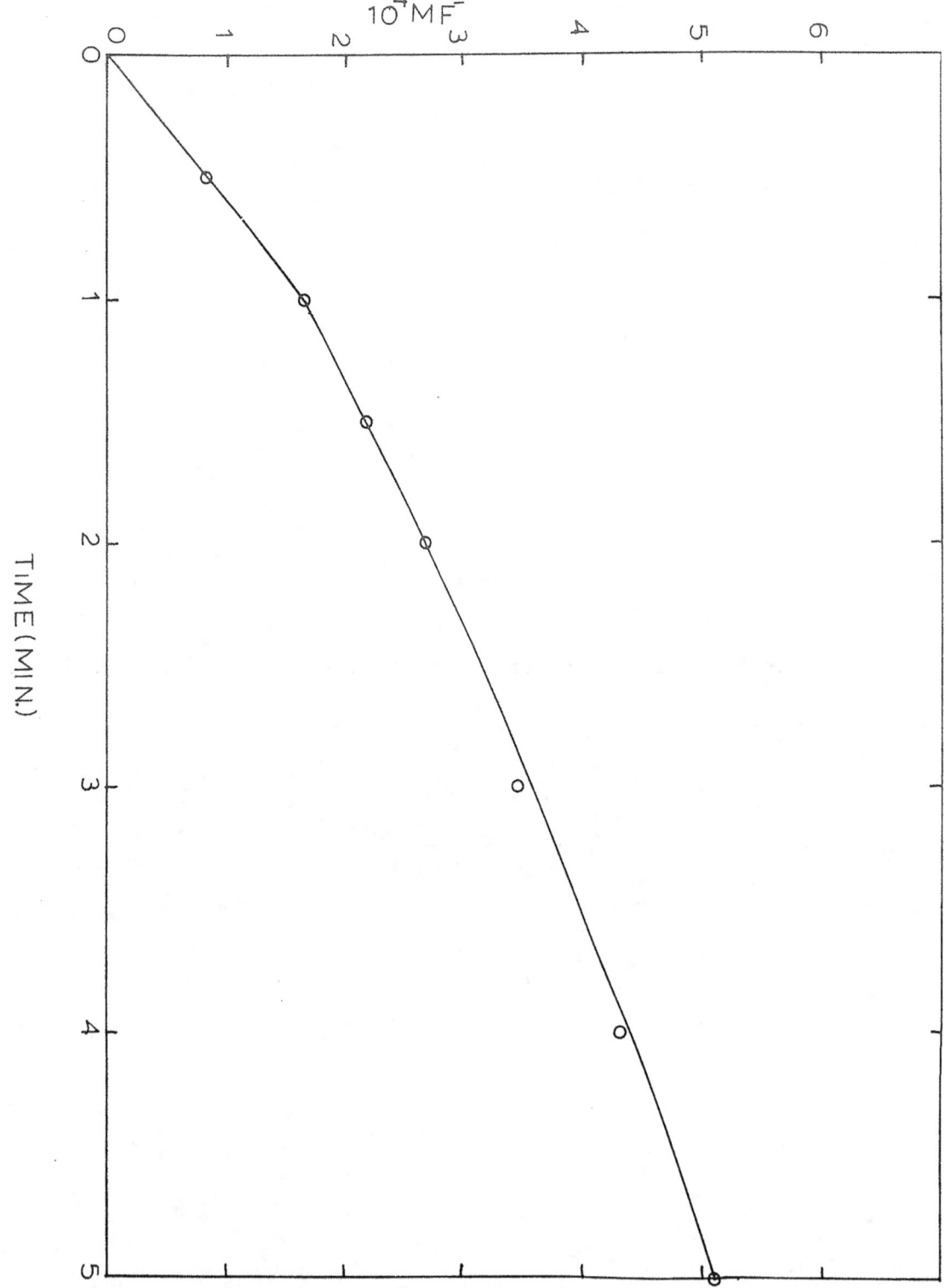

Figure 7

Quantum yield of fluoride as a function of $[SF_6]$. dose = 8 x 10^{19} quanta for $[SF_6] >$ 2.4 x 10^{-4}M and for $[SF_6] \leqslant$ 2.4 x 10^{-4}M, dose = 1.08 x 10^{20} quanta

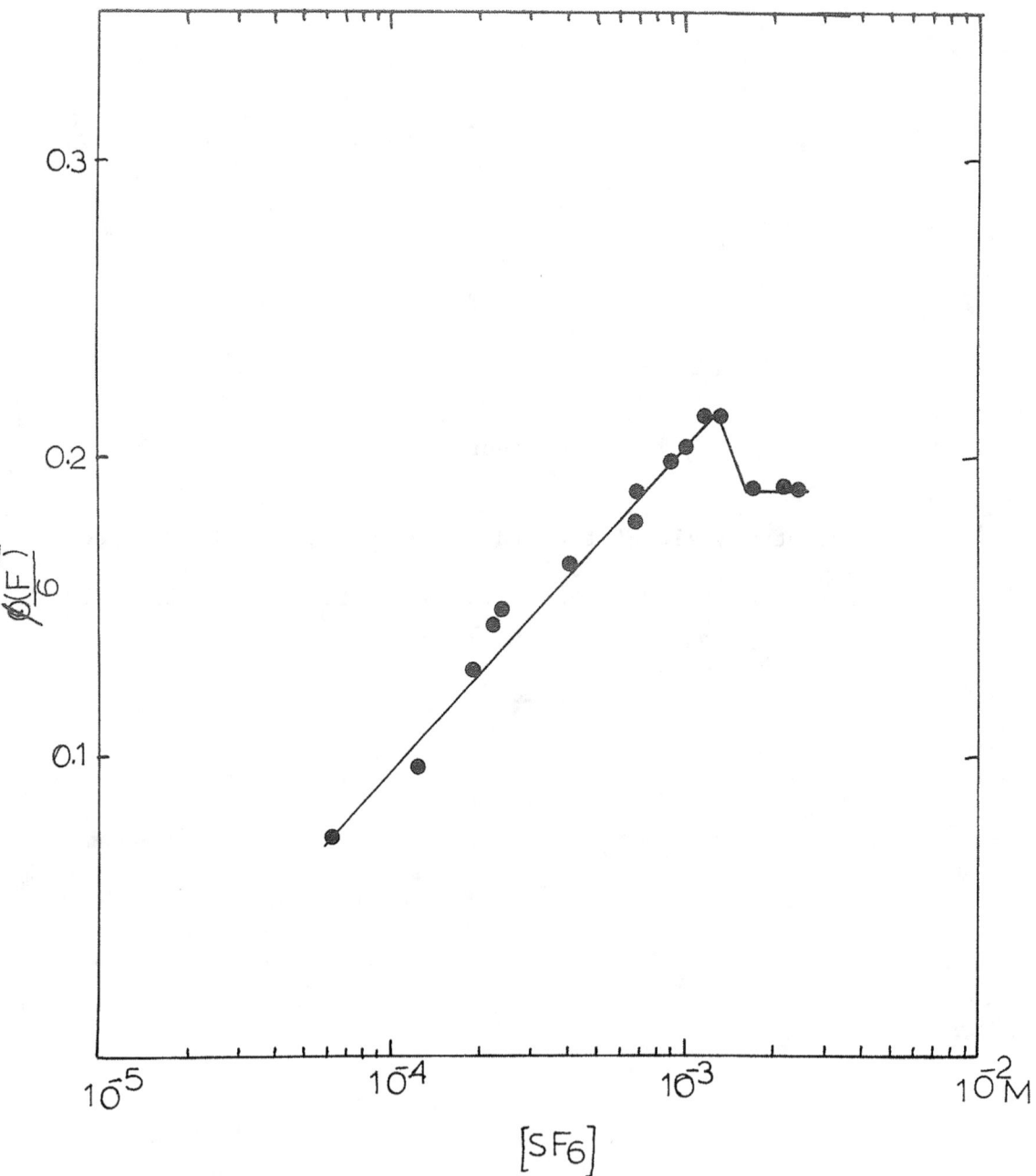

Figure 8

Yield of fluoride as a function of irradiation time. $[SF_6]$= 1.2 x 10^{-3}M, dose rate = 8 x 10^{19} quanta /min.

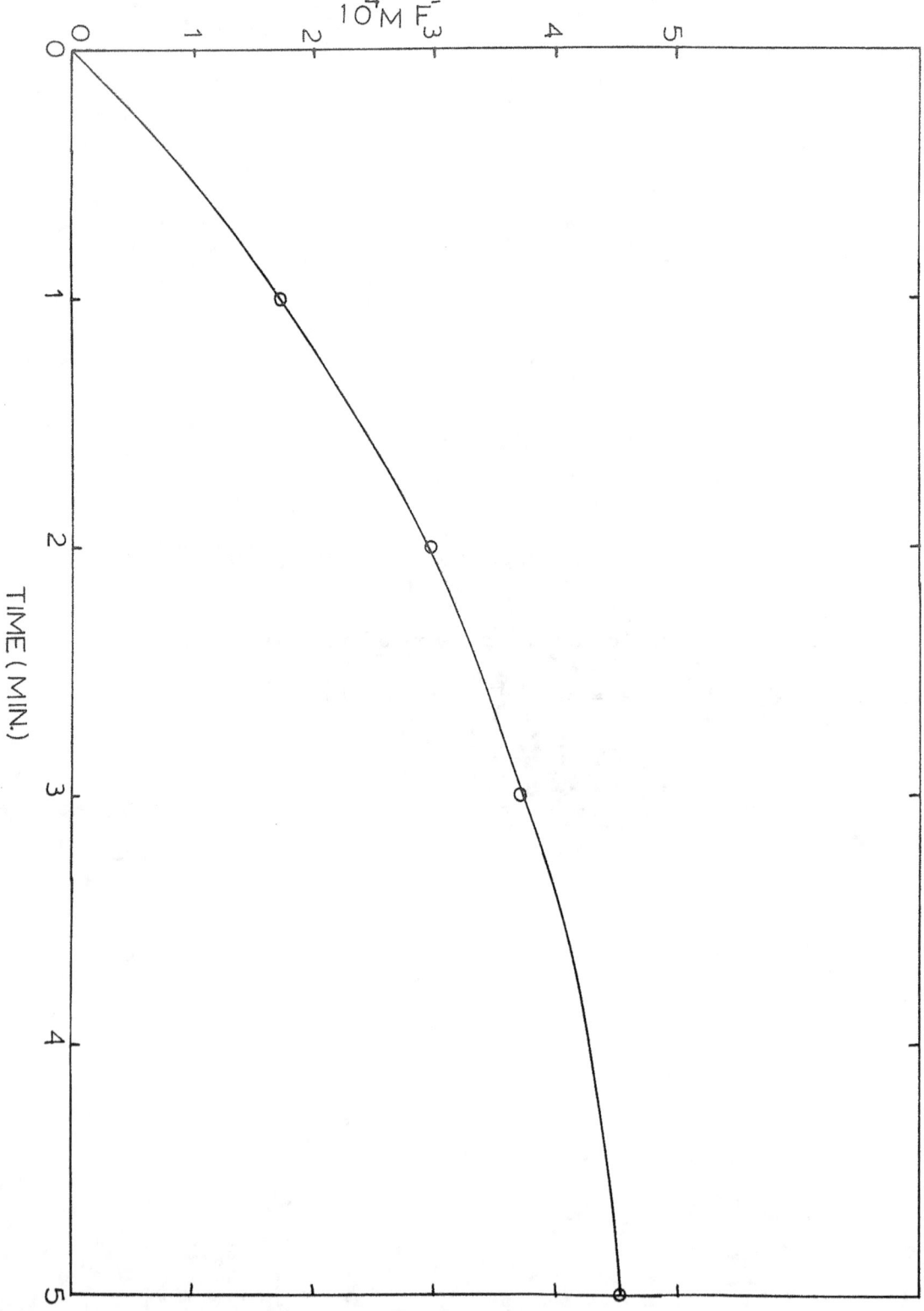

3.3.1

Yield versus dose plot (fig. 8) was attempted at 1.2×10^{-3}M SF$_6$ which gave maximum quantum yield of fluoride in fig. 7. Fig. 8 shows the continuously curving fluoride yield and from the initial slope $\frac{\Phi(F^-)}{6} = 0.225 \pm 0.005$. Negligible amount of iodine was formed in all studies with KI and SF$_6$.

3.3.2

In preliminary experiments with 5×10^{-3}M solution of KI containing 2.4×10^{-4}M SF$_6$ the quantum yield of fluoride was much higher than reported in fig. 6. More thorough degassing of KI solution (before the addition of SF$_6$) resulted in the reproducible yields of fluoride (fig. 6). The high fluoride yield in the preliminary experiments may be due to effects of residual oxygen remaining due to incomplete degassing of the KI solution. The same trend of high fluoride yield was noted in the competition experiments with N$_2$O and SF$_6$.

3.4

Competition studies: KI - SF$_6$ - N$_2$O Systems

The following equations can be derived for the competition reactions of hydrated electrons with SF$_6$ and N$_2$O: (For the sequence of reactions in the case of iodide system, see Mechanism I and II in the discussion, in Chapter V) The results of the competition plots for each system are summarised in table 1.

$$[\Phi(N_2)]^{-1} = [\Phi^o(\bar{e}_{aq})]^{-1} + [\Phi^o(\bar{e}_{aq})]^{-1} \cdot \frac{k_{e\,+\,SF_6}[SF_6]}{k_{e\,+\,N_2O}[N_2O]} \quad \ldots \ldots \quad (9)$$

$$[\frac{\Phi(F^-)}{6}]^{-1} = [\Phi^o(\bar{e}_{aq})]^{-1} + [\Phi^o(\bar{e}_{aq})]^{-1} \cdot \frac{k_{e\,+\,N_2O}[N_2O]}{k_{e\,+\,SF_6}[SF_6]} \quad \ldots \ldots \quad (10)$$

$$[\Phi(I_2)]^{-1} = [\Phi^o(\bar{e}_{aq})]^{-1} + [\Phi^o(\bar{e}_{aq})]^{-1} \cdot \frac{k_{e\,+\,SF_6}[SF_6]}{k_{e\,+\,N_2O}[N_2O]} \quad \ldots \ldots \quad (11)$$

Where $\Phi(N_2)$, $\Phi(F^-)$ and $\Phi(I_2)$ are the measured yields of the products of reactions (12 - 16) and (29 - 33) and $\Phi^o(\bar{e}_{aq})$ is the total measured yield of hydrated electrons.

A sample derivation of equation (9) from the reactions is given below and equations (10) and (11) can be derived in a similar manner:

$$\Phi(N_2) = \Phi^o(\bar{e}_{aq}) \frac{k_{13}(\bar{e}_{aq})(N_2O)}{k_{13}(\bar{e}_{aq})(N_2O + k_{29}(\bar{e}_{aq})(SF_6)}$$

Rearranging

$$\Phi(N_2) = \Phi^o(\bar{e}_{aq}) \cdot \frac{1}{1 + \frac{k_{29}(SF_6)}{k_{13}(N_2O)}}$$

OR

$$[\Phi(N_2)]^{-1} = [\bar\Phi^{0}(\bar e_{aq})]^{-1} + [\bar\Phi^{0}(\bar e_{aq})]^{-1} \frac{k_{29}(SF_6)}{k_{13}(N_2O)}$$

Thus the plots of $[\bar\Phi(N_2)]^{-1}$ vs $(N_2O)^{-1}$; $[\Phi(F^-)]^{-1}$ vs $(SF_6)^{-1}$ and $[\Phi(I_2)]^{-1}$ vs $(N_2O)^{-1}$ should give straight lines with the intercepts equal to $\lfloor\bar\Phi^{0}(\bar e_{aq})\rfloor^{-1}$ and slopes equal to

$$\frac{k_{29}(SF_6)}{k_{13}} \quad , \qquad \frac{k_{13}(N_2O)}{k_{29}} \quad , \quad \text{and} \quad \frac{k_{29}(SF_6)}{k_{13}} \quad \text{respectively.}$$

So, from the competition studies of hydrated electrons with SF_6 and N_2O, it is possible to determine the total measured yield of hydrated electrons and the rate constant ratio $\frac{k_{29}}{k_{13}}$. Obviously, one must obtain the same rate constant ratio and the yield of $\bar\Phi^{0}(\bar e_{aq})$ by treating the kinetic data using any of the equations (9, 10 or 11) if the kinetic reactions postulated in the mechanisms (I) - (III) are true.

3.4.1

Kinetic analysis of fluoride yield:

3.4.1.1

Irradiation time = 1 minute, vary $[N_2O]$. The non-degassed and de-gassed fluoride yields are plotted in figs. 9 and 10 (tables 2 and 3) respectively. Due to some scatter in the data, the intercept value has been fixed in both cases and $\frac{\Phi(F^-)}{6} = 0.225 \pm 0.01$, from the intercept

Figure	$\dfrac{k_{29}}{k_{13}}$	$\Phi \dfrac{(F^-)}{6}$	Figure	$\dfrac{k_{29}}{k_{13}}$	$\Phi(N_2)$
(Kinetic analysis of F^- yield)			(Kinetic analysis of N_2 yield)		
9	4.7 ± 0.15	0.225 ± 0.01	17	19.7 ± 1	0.245 ± 0.01
10	4.08 ± 0.15	0.225 ± 0.01			
11[*]	3.9 ± 0.15	0.238 ± 0.01	18	21.5 ± 1	0.23 ± 0.01
12	3.5 ± 0.15	0.238 ± 0.01			
13(1)[*]	4.8 ± 0.15	0.238 ± 0.01	19(1)	18 ± 0.6	0.24 ± 0.005
13(2)[*]	5.3 ± 0.15	0.238 ± 0.01			
14(1)	4.2 ± 0.15	0.238 ± 0.01	19(2)	20 ± 0.5	0.24 ± 0.005
14(2)	4 ± 0.15	0.238 ± 0.01			
15[*]	3.9 ± 0.15	0.238 ± 0.01			
16[*]					
($Fe(CN)_6^{4-}$ system)	1.8 ± 0.15	1.05 ± 0.1			

(Figures marked with * represent non - degassed fluoride yields and without * represent degassed fluoride yields.)

Table I

KI - SF_6 - N_2O SYSTEM

$[SF_6] = 1.58 \times 10^{-4}$ M, irradiation time = 1 min.

$[N_2O] \times 10^3$ M	$\frac{\Phi(F^-)}{6}$ (Non-degassed)	$\Phi(I_2)$ (Non-degassed)
1.73	0.073	0.12
1.9	0.065	0.12
3.05	0.045	0.17
3.43	0.04	0.16
4.19	0.036	0.18
5.4	0.028	0.15
6.17	0.023	0.2
6.65	0.022	0.19
8.6	0.02	0.19

Table II

KI - SF$_6$ - N$_2$O SYSTEM

$$[SF_6] = 1.58 \times 10^{-4} M, \text{ irradiation time} = 1 \text{ min.,}$$

$[N_2O] \times 10^3$ M	$\Phi(N_2)$	$\dfrac{\Phi(F^-)}{6}$ (degassed)	$\Phi(N_2) + \dfrac{\Phi(F^-)}{6}$	$\Phi(I_2)$ (degassed)
1.73	0.09	0.073	0.16	0.098
1.87	0.1	0.065	0.16	0.10
2.5	0.1	0.05	0.16	0.12
3.5	0.126	0.037	0.16	0.135
5.23	0.15	0.022	0.17	0.137
7.03	0.17	0.019	0.19	0.137
8.6	0.18	0.018	0.20	0.140

Table III

Figure 9

Non - degassed $[\Phi(F^-)]^{-1}$ as a function of $[N_2O]$. $[SF_6] =$ 1.58 x 10^{-4}M, dose = 1.08 x 10^{20} quanta.

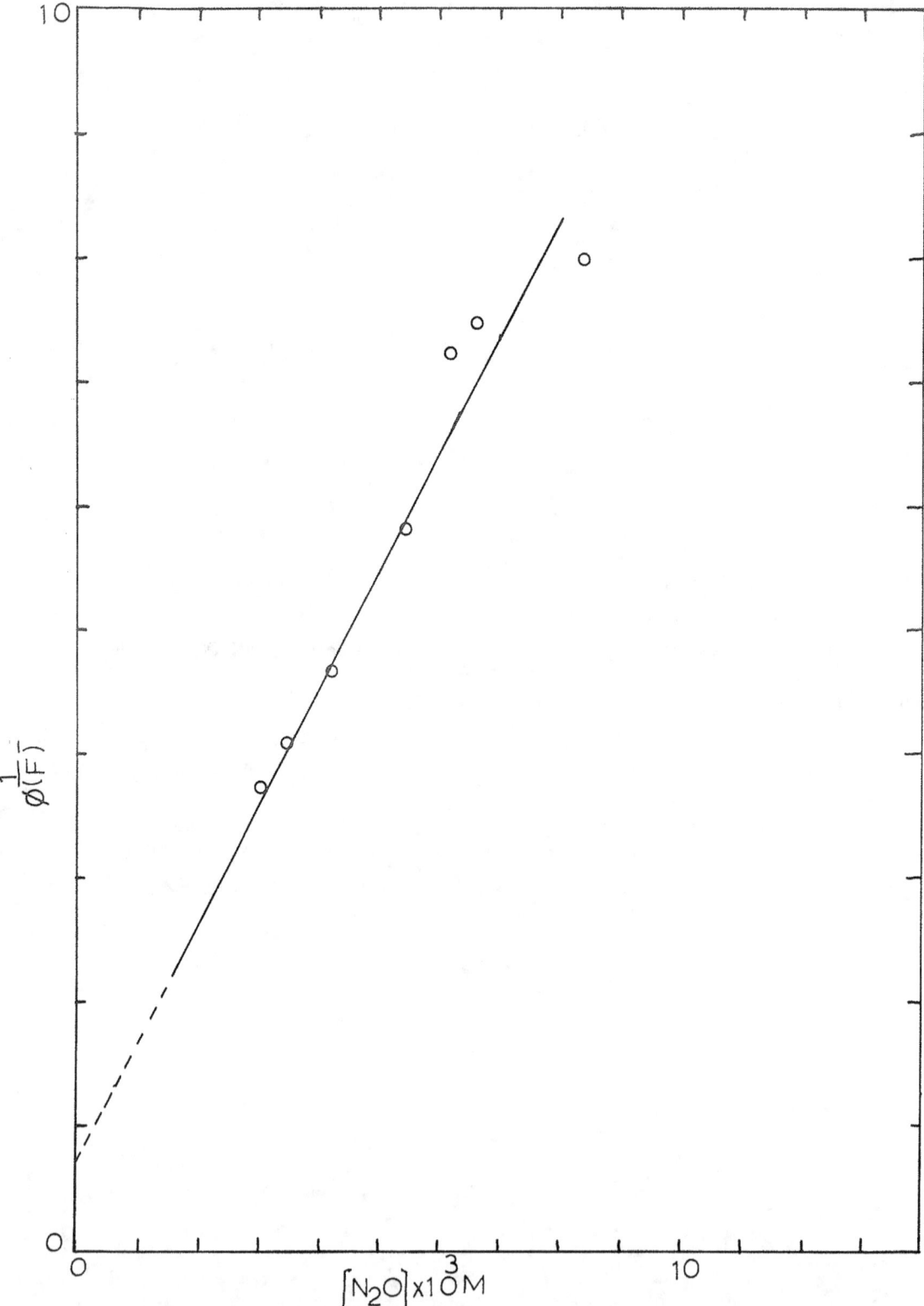

Figure 10

Degassed $[\dot{\Phi}(F^-)]^{-1}$ as a function of $[N_2O]$. $[SF_6] = 1.58 \times 10^{-4} M$, dose $= 1.08 \times 10^{20}$ quanta.

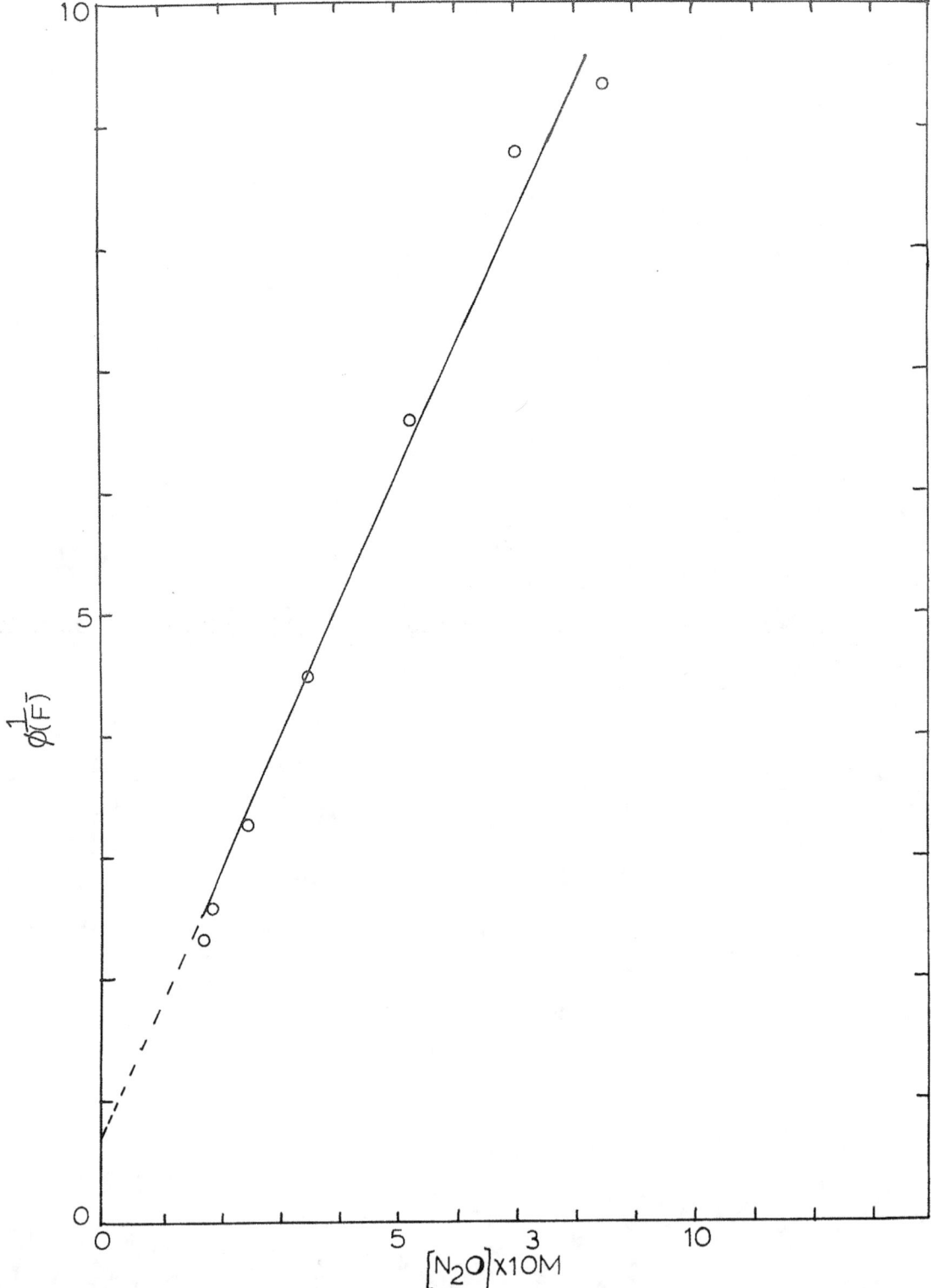

values, 0.7 and 0.75 in figs. 9 and 10 respectively, was obtained. The rate constant ratio $\dfrac{k_{e + SF_6}}{k_{e + N_2O}}$ is 4.7 ± 0.15 in fig. 9 and 4.08 ± 0.15 in fig. 10. The difference in fluoride yields in both the non-degassed and degassed cases, at the same scavenger concentration, is within the limits of experimental error but it was noted from the potential measurements of the fluoride yields that their values were higher in the case of the degassed one and hence fluoride yield was lower than the non-degassed, at the same scavenger concentration.

3.4.1.2

$[SF_6] = 1.58 \times 10^{-4} M$

Irradiation time $= 2$ minutes, vary $[N_2O]$. The non-degassed and degassed quantum yields of fluoride are plotted in figs. 11 and 12 (Tables 4 and 5). The intercept value is fixed at 0.7 in both cases and hence $\dfrac{\phi(F^-)}{6} = 0.238 \pm 0.1$.

From the slopes the rate constant ratio $\dfrac{k_{e + SF_6}}{k_{e + N_2O}}$ in figs. 11 and 12 are 3.9 ± 0.15 and 3.5 ± 0.15 respectively. The same behaviour in fluoride yield that it decreases on degassing as observed with irradiation time $= 1$ minute, (figs. 9 and 10) is more apparent with 2 minutes of irradiation time.

3.4.1.3

$[N_2O] = 1.73 \times 10^{-3} M$ and $3.46 \times 10^{-3} M$

Irradiation time $= 1$ minute, vary $[SF_6]$. The non-degassed and degassed

KI - SF$_6$ - N$_2$O SYSTEM

$[SF_6] = 1.58 \times 10^{-4}$ M, irradiation time = 2 min.,

[N$_2$O] x 10^{3}M	Φ(N$_2$)	Φ(I$_2$) (degassed)	[N$_2$O] x 10^{3}M	$\Phi\frac{(F^-)}{6}$ (degassed)	$\Phi(N_2)+\Phi\frac{(F^-)}{6}$
1.73	0.074	0.098			
2.28	0.095	0.167	2.1	0.044	0.139
2.6	0.097	0.15			
2.94	0.106	0.11	2.87	0.037	0.143
3.15	0.167	0.14			
3.26	0.16	0.156	3.29	0.033	0.193
3.5	0.15	0.164			
3.67	0.14	0.158	3.71	0.032	0.172
4.02	0.14	0.16			
4.37	0.12	0.154			
4.79	0.123	0.154	4.75	0.022	0.145
4.85	0.13	0.154			
6.55	0.134	0.156	6.52	0.016	0.150
7.76	0.177	0.17			
8.66	0.173	0.164	8.9	0.015	0.188

Table IV

KI - SF₆ - N₂O SYSTEM

$[SF_6] = 1.58 \times 10^{-4}M$, irradiation time = 2 min.,

$[N_2O] \times 10^3 M$	$\Phi(F^-)$ (non-degassed)	$\Phi(I_2)$ (non-degassed)
1.84	0.071	0.126
2.87	0.035	0.153
3.5	0.028	0.135
4.57	0.024	0.157
6.45	0.021	0.139
8.7	0.017	0.173

Table V

Figure 11

Non - degassed $[\Phi(F^-)]^{-1}$ as a function of $[N_2O]$. $[SF_6] = 1.58$ x 10^{-4}M, dose = 2.16 x 10^{20} quanta.

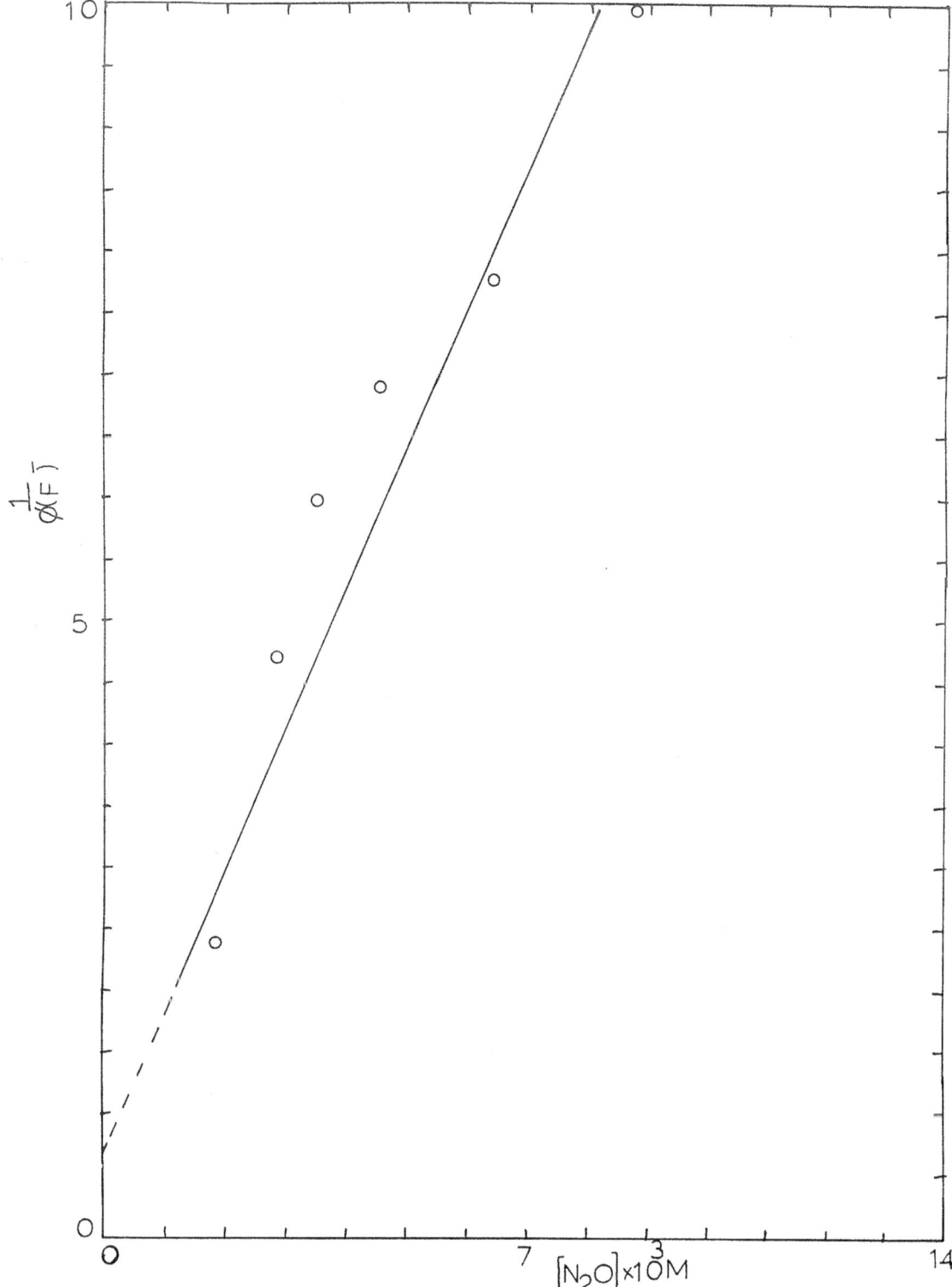

Figure 12

Degassed $[\Phi(F^-)]^{-1}$ as a function of $[N_2O]$. $[SF_6] = 1.58 \times 10^{-4}$ M, dose $= 2.16 \times 10^{20}$ quanta.

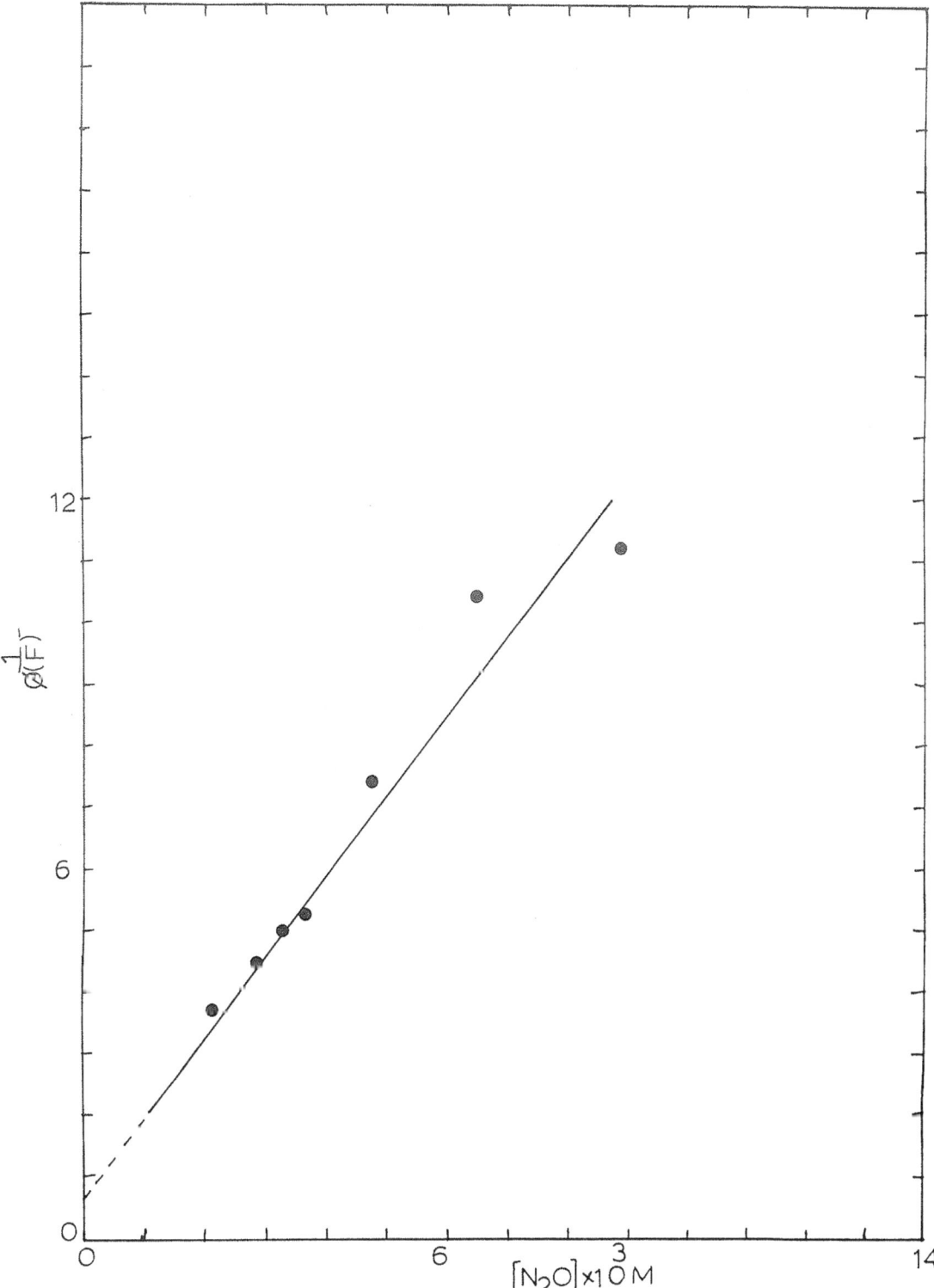

fluoride yields are represented in figs. 13 (I and II) and 14(I and II) and in tables 6 and 7. The intercept value $\frac{\Phi(F^-)}{6} = 0.238 \pm 0.01$ was fixed in both the figures and the calculated rate constant ratio was

$$\frac{k_{e + SF_6}}{k_{e + N_2O}} = 4.8 \pm 0.15$$

when $[N_2O] = 3.46 \times 10^{-3}M$ (fig. 13 (I) and 5.3 ± 0.015 when N_2O concentration $= 1.73 \times 10^{-3}M$ (fig. 13(II) and, in the case of degassed fluoride yields, these values are 4.2 ± 0.15 and 4 ± 0.15 respectively (figs. 14(I) and (II). Again, the effect on the fluoride yields due to degassing is apparent (tables 6 and 7).

3.4.1.4

Vary $[N_2O]$ and $[SF_6]$ at high pressures of SF_6, Irradiation time = 1 minute.

The variation of the quantum yield of fluoride with the ratio of $[\frac{N_2O}{SF_6}]$ is represented in fig. 15. The rate constant ratio $\frac{k_{e + SF_6}}{k_{e + N_2O}} = $

3.9 ± 0.15, from the line drawn by fixing the intercept value $\frac{\Phi(F^-)}{6} = $ 0.238 ± 0.01, is lower than determined at low pressures of SF_6(figs. 9, 13(I and II)) in the non-degassed manner. The fluoride yield increases by increasing $[SF_6]$ or decreasing $[N_2O]$ (table 8).

3.4.1.5

$Fe(CN)_6^{4-}$ - SF_6 - N_2O System: Vary both $[SF_6]$ and $[N_2O]$, irradiation time = 1 minute:

KI – SF$_6$ – N$_2$O SYSTEM

$[N_2O] = 1.73 \times 10^{-3}M$; irradiation time = 1 min.,

$[SF_6] \times 10^4 M$	$\Phi(N_2)$	$\dfrac{\Phi(F^-)}{6}$ (degassed)	$\Phi(N_2)+\dfrac{\Phi(F^-)}{6}$	$\Phi(I_2)$ (degassed)
0.51	0.15	0.023	0.17	0.165
1.08	0.11	0.059	0.17	0.148
1.58	0.088	0.072	0.16	0.098
2.22	0.07	0.074	0.14	0.106

$[N_2O] = 3.46 \times 10^{-3}M$; irradiation time = 1 min.,

0.51	0.185	0.0123	0.2	0.185
1.08	0.153	0.023	0.176	0.196
1.58	0.126	0.037	0.163	0.135
1.9	0.113	–	–	0.175

Table VI

$$\underline{KI - \underline{SF}_6 - \underline{N_2O} \text{ SYSTEM}}$$

$$[\underline{N_2O}]= 1.73 \times 10^{-3}M; \text{ irradiation time = 1 min.,}$$

$[SF_6] \times 10^4 M$	$\Phi_6 (F^-)$ (non-degassed)	$\Phi(I_2)$ (non-degassed)
0.51	0.0297	0.174
1.08	0.065	0.124
1.58	0.0725	0.122
2.22	0.082	0.112

$$[\underline{N_2O}]= 3.46 \times 10^{-3}M; \text{ irradiation time = 1 min.,}$$

0.51	0.013	0.197
1.08	0.033	0.183
1.58	0.041	0.160
2.06	0.056	0.167

Table VII

KI - SF$_6$ - N$_2$O SYSTEM

Irradiation time = 1 min., vary [N$_2$O] and [SF$_6$]

[N$_2$O] x 10^3 M	[SF$_6$] x 10^3 M	$\dfrac{[N_2O]}{SF_6}$	$\dfrac{\Phi(F^-)}{6}$ (non-degassed)	$\Phi(I_2^-)$
22.7	0.998	22.75	0.033	0.28
20.65	1.39	14.8	0.051	0.23
6.93	0.69	10.03	0.071	0.23
6.93	0.99	6.99	0.077	0.21
4.16	0.86	4.84	0.106	0.14
2.53	1.1	2.26	0.14	0.055
2.46	1.77	1.39	0.18	0.066
2.46	0.705	3.49	0.12	0.11
2.357	0.41	5.7	0.09	0.103

Table VIII

Figure 13

Non - degassed $[\Phi(F^-)]^{-1}$ as a function of $[SF_6]^{-1}$. $[N_2O] =$ 3.46 x 10^{-3}M, dose = 1.08 x 10^{20} quanta (Figure 13 (I)). $[N_2O] =$ 1.73 x 10^{-3}M, dose = 1.08 x 10^{20} quanta (Figure 13 (II)).

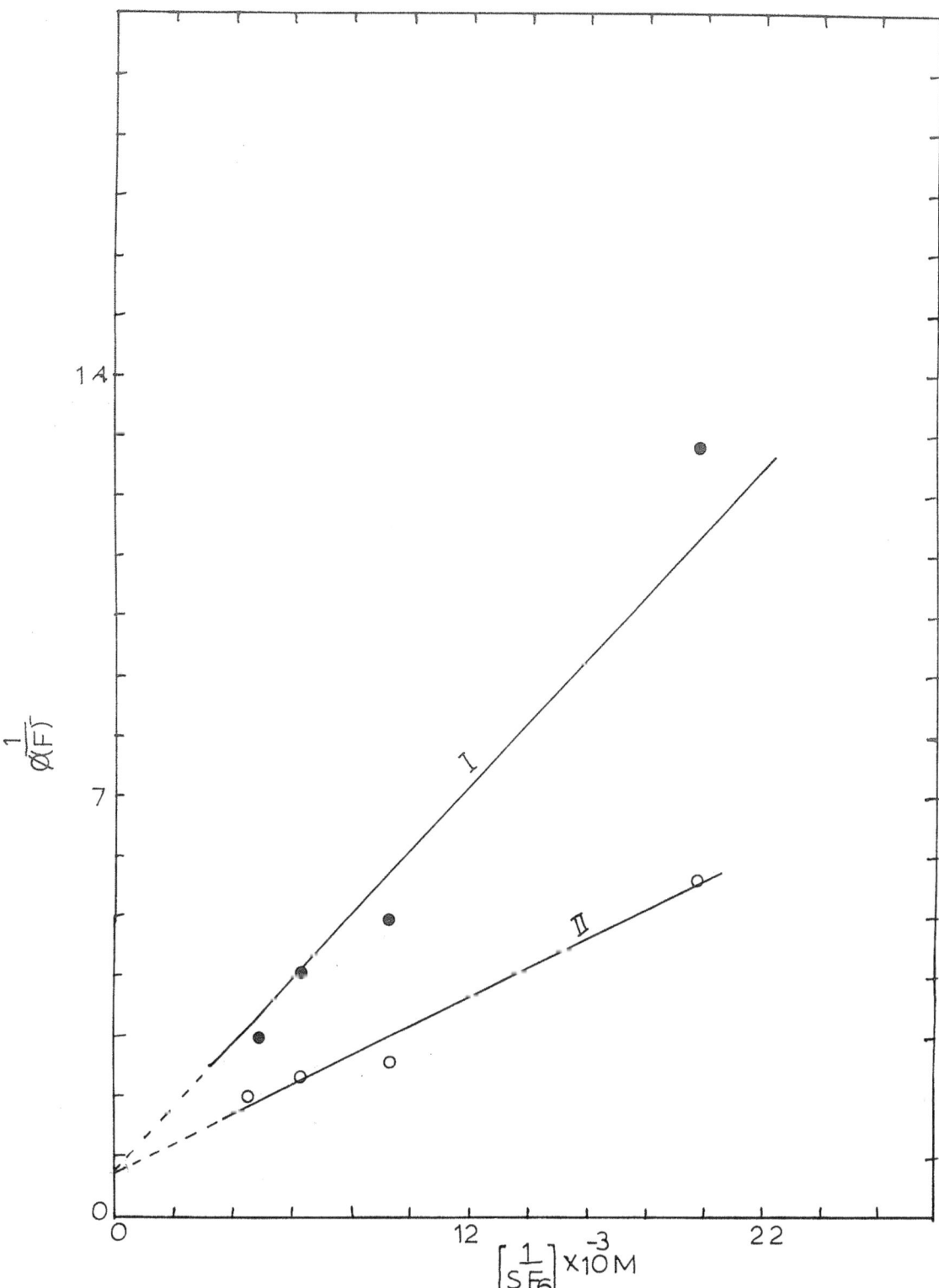

Figure 14

Degassed $[\Phi(F^-)]^{-1}$ as a function of $[SF_6]^{-1}$. $[N_2O] = 3.46$ x 10^{-3}M, dose = 1.08 x 10^{20} quanta (Figure 14 (I)). $[N_2O] = 1.73$ x 10^{-3}M, dose = 1.08 x 10^{20} quanta (Figure 14(II)).

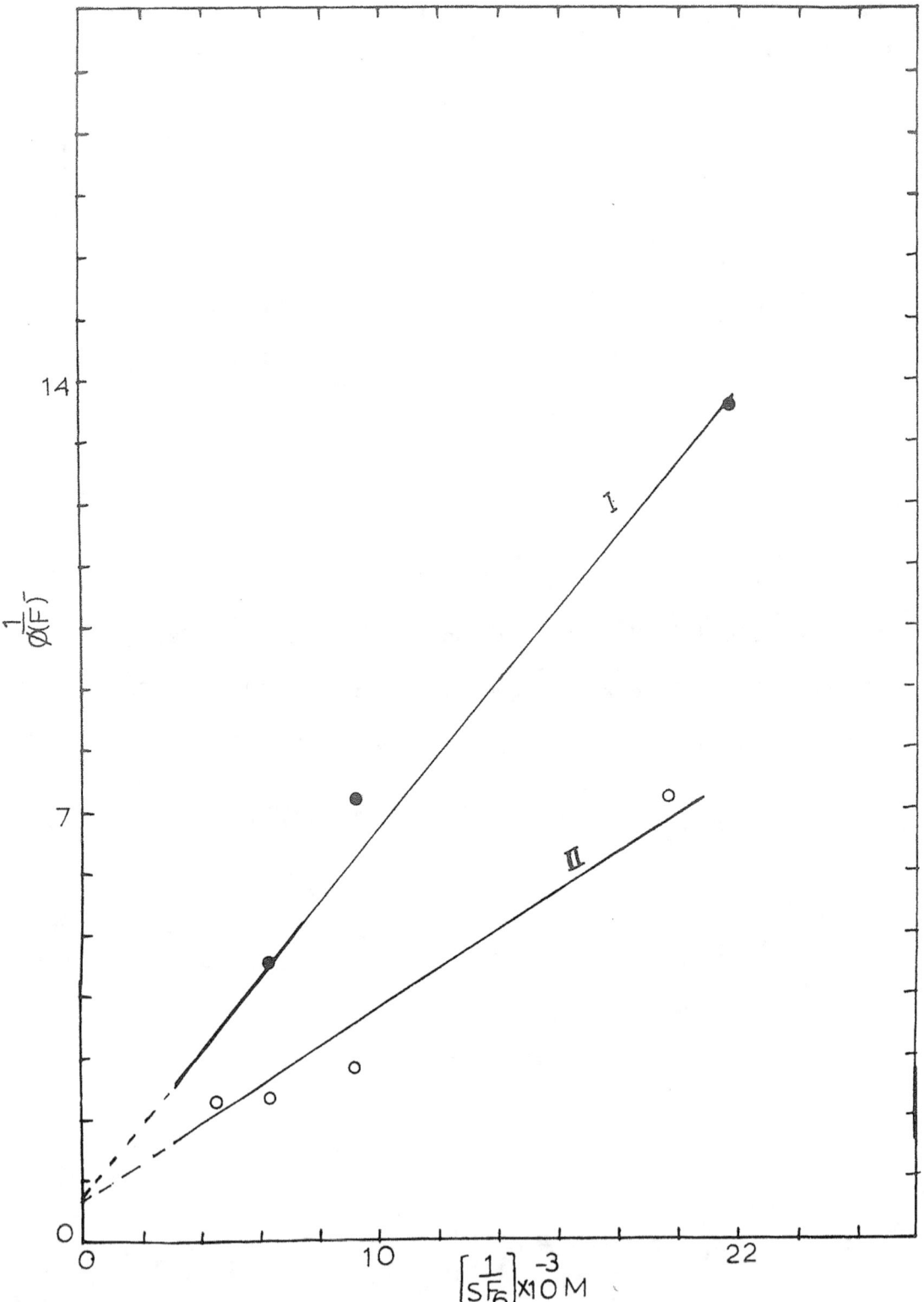

Figure 15

Non - degassed $[\Phi(F^-)]^{-1}$ as a function of $[\frac{N_2O}{SF_6}]$. dose = 8 x 10^{19} quanta

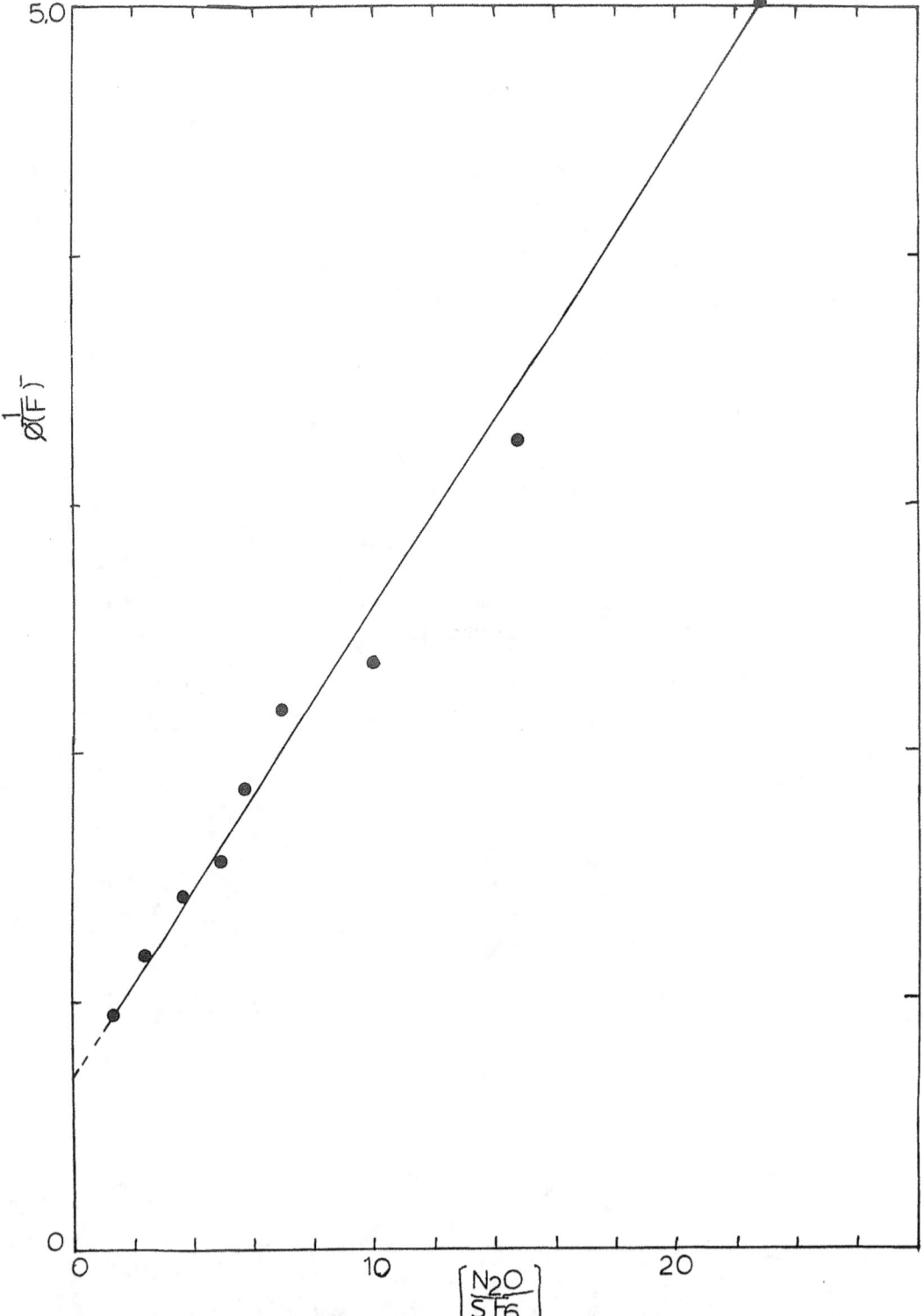

Figure 16

Non - degassed $[\Phi(F^-)]^{-1}$ as a function of $[\frac{N_2O}{SF_6}]$ in the $Fe(CN)_6^{4-}$ system. dose $= 8 \times 10^{19}$ quanta

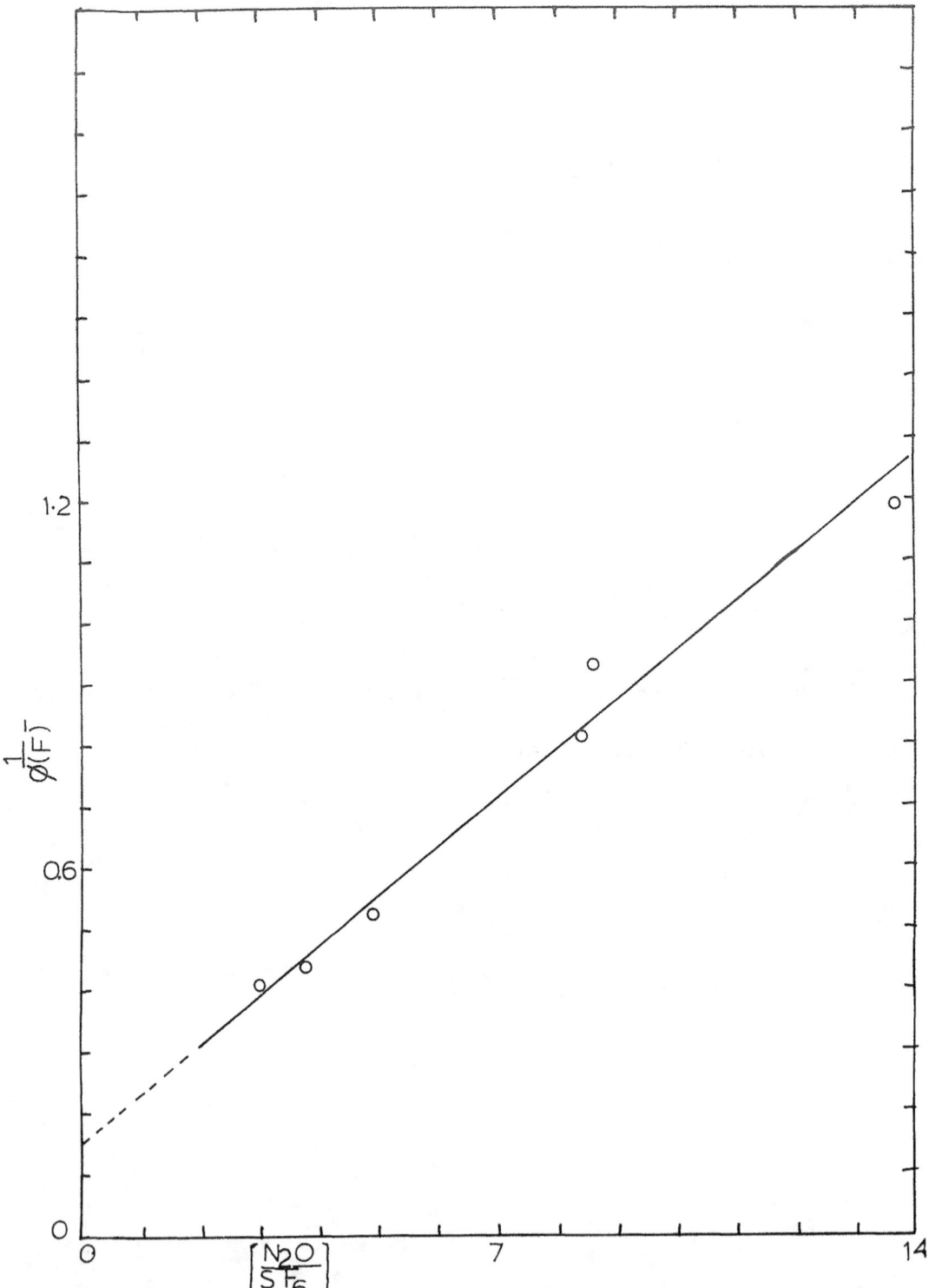

Competition experiments were done with 5×10^{-3}M aqueous solution of $Fe(CN)_6^{4-}$ using $2537A^{\circ}$ light. The studies were carried out at high pressure of SF_6 and the results are shown in fig. 16. There is some scatter in the data; however, the best intercept value obtained from the plot was 0.15 which gave $\Phi \frac{(F^-)}{6} = 1.05 \pm 0.1$ and the rate constant ratio,

$$\frac{k_{e + SF_6}}{k_{e + N_2O}} = 1.8 \pm 0.15$$ which is in good agreement with the value 1.67 obtained by Asmus and Fendler (60) in their radiation-chemical studies of this system. The quantum yield of $Fe(CN)_6^{3-}$ in all these experiments was constant $\Phi(Fe(CN)_6^{3-}) = 1.84 \pm 0.05$.

3.4.2

Kinetic analysis of nitrogen yield:

The yield of nitrogen was zero in the first freeze-pump cycle if technique 2.9.1.a was used for its measurement, and very little nitrogen was formed in the second and third cycles. If liquid nitrogen is replaced by a dry ice + acetone bath in the second and third cycles, the yield of nitrogen is approximately 10% less than formed in the three cycles if only dry ice + acetone is used as the refrigerant of water (technique 2.9.1.b). Only a few experiments were tried to see the effects of liquid nitrogen and dry ice + acetone.

In all the following experiments for the analysis of nitrogen, technique 2.9.1.b was used. The kinetic analysis of nitrogen yield has been attempted in the same manner as in the case of the fluoride yield but the rate constant ratio $\dfrac{k_{e + SF_6}}{k_{e + N_2O}}$ is much different in the former than what was obtained in the latter case.

3.4.2.1

$[SF_6]$ = 1.58 x 10^{-4}M, Irradiation Time = 1 minute. Vary $[N_2O]$.
From the results represented in fig. 17 (Table III),

$$\frac{k_{e + SF_6}}{k_{e + N_2O}} = 19.7 \pm 1 \text{ and the intercept value } \Phi(N_2) = 0.245 \pm 0.01$$

3.4.2.2

$[SF_6]$ = 1.58 x 10^{-4}M, Irradiation Time = 2 minutes, Vary $[N_2O]$ concentration.

At about 3.2 x 10^{-3}M $[N_2O]$, the nitrogen yield is unusually high
but the corresponding fluoride yields (degassed or non-degasses) do not
show the reverse behaviour, i.e. they may be low if the nitrogen is high
at particular concentrations of N_2O and SF_6. The results are shown in
fig. 18 (table 4) from which the rate constant ratio $\dfrac{k_{e + SF_6}}{k_{e + N_2O}}$ =

21.5 $\pm$ 1 and the intercept value $\Phi(N_2)$ = 0.23 $\pm$ 0.01.

3.4.2.3

$[N_2O]$ = 1.73 x 10^{-3}M and 3.46 x 10^{-3}M, Irradiation time = 1 minute,
vary $[SF_6]$.

The results are represented in fig. 19 (I) and (II) (table 6), from
which the rate constant ratio $\dfrac{k_{e + SF_6}}{k_{e + N_2O}}$ = 18 $\pm$ 0.6 (fig. 19, (I)), and

$$\frac{k_{e + SF_6}}{k_{e + N_2O}} = 20 \pm 0.5 \text{ (fig. 19, (II)), and the common intercept value}$$

Figure 17

$[\Phi(N_2)]^{-1}$ as a function of $[N_2O]$. $[SF_6] = 1.58 \times 10^{-4}$ M, dose $= 1.08 \times 10^{20}$ quanta.

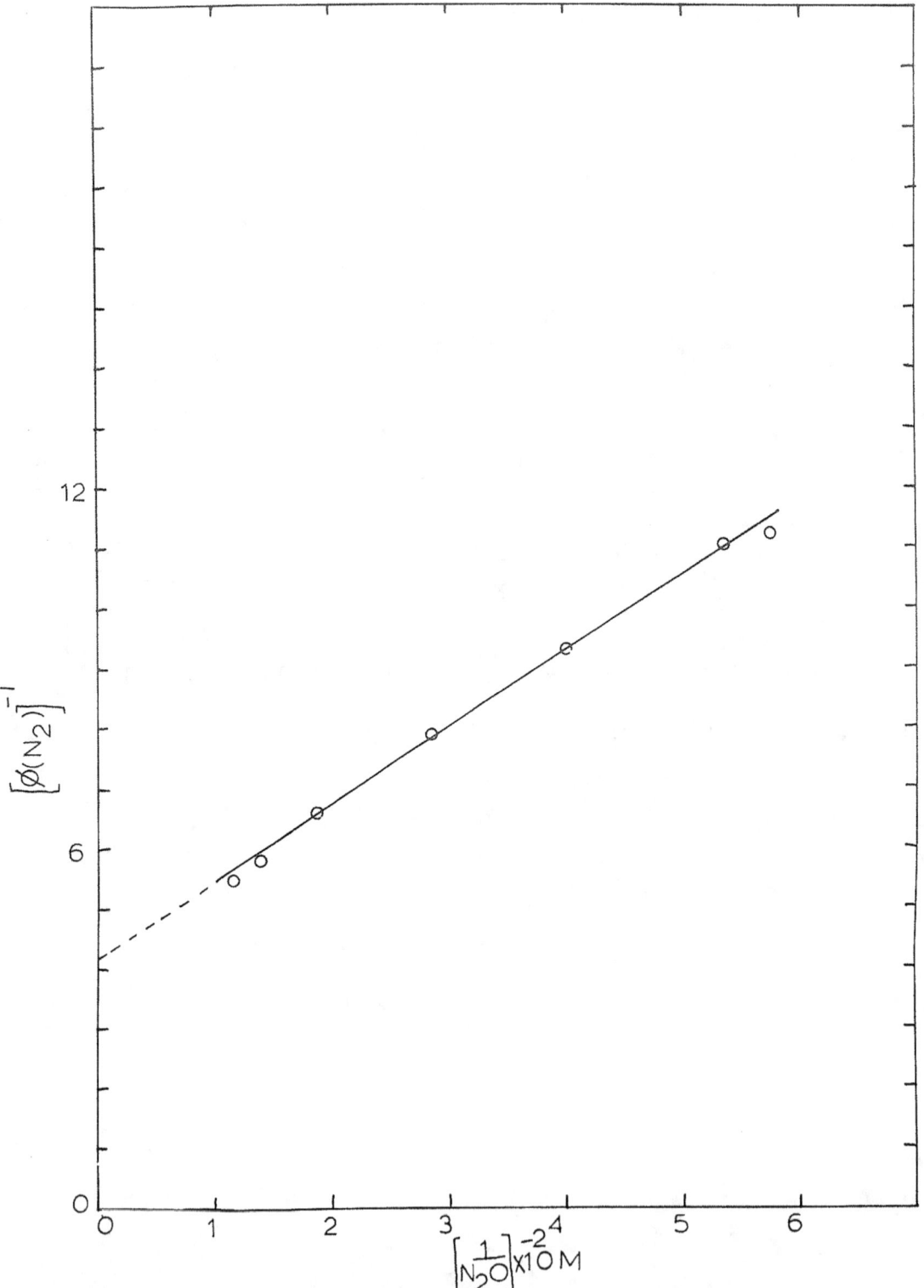

Figure 18

$[\Phi(N_2]^{-1}$ as a function of $[N_2O]^{-1}$. $[SF_6] = 1.58 \times 10^{-4}M$, dose $= 2.16 \times 10^{20}$ quanta.

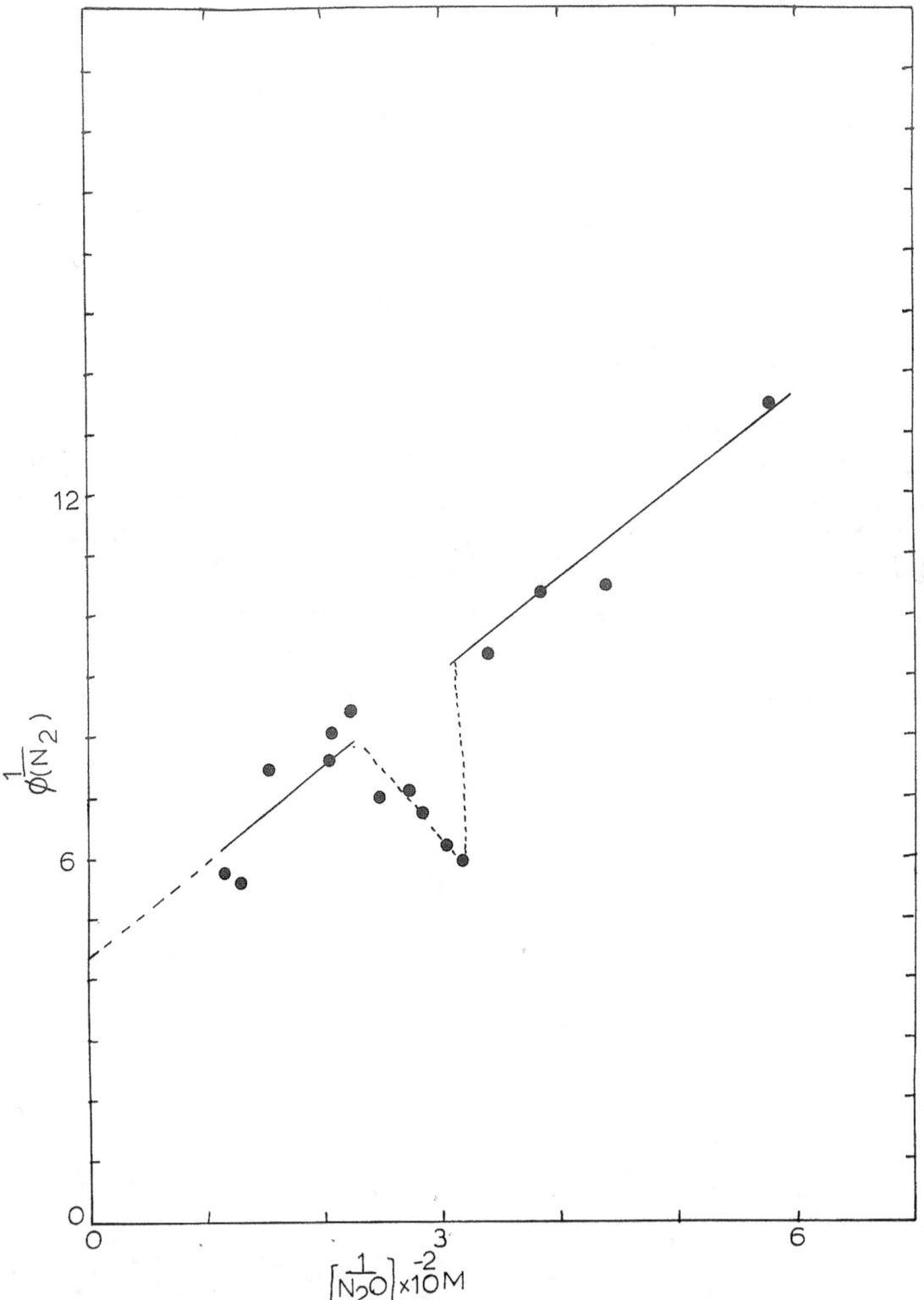

Figure 19

$[\Phi(N_2)]^{-1}$ as a function of $[SF_6]$. $[N_2O] = 1.73 \times 10^{-3}M$, dose $= 1.08 \times 10^{20}$ quanta (Figure 19(I)). $[N_2O] = 3.46 \times 10^{-3}M$, dose $= 1.08 \times 10^{20}$ quanta (Figure 19 (II)).

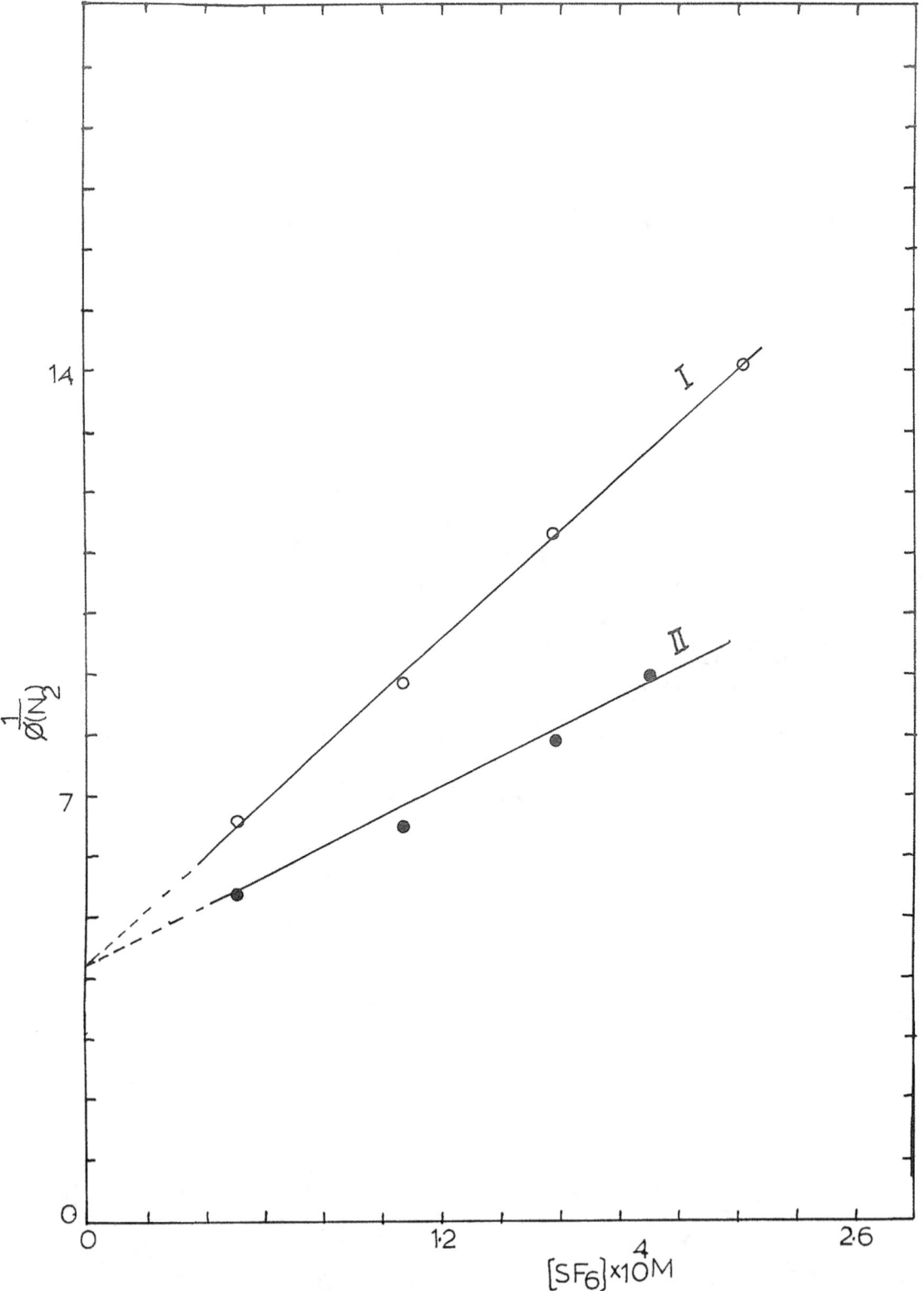

19

$\Phi(N_2) = 0.24 \pm 0.005.$

3.4.3

Kinetic analysis of Iodine yield:

The kinetic analysis of the iodine yields (tables 2 - 8) measured from the degassed and non-degassed solutions was attempted in the same manner as for the fluoride and nitrogen yield analyses. The plots obtained were not meaningful due to the scatter in the experimental data; however, the accuracy of the measurement of iodine yield is within $\pm$ 2% .

Tables 2 - 7 indicate that neither the degassed nor the non-degassed iodine yield is equal to the nitrogen yield. The summation of either the non-degassed or the degassed fluoride and nitrogen yields at approximately the same scavenger concentrations is not constant. Table 8 shows that, by increasing the $[N_2O]$ and decreasing the $[SF_6]$ the iodine yield increases beyond the maximum quantum yield of 0.23 obtained only with $2.6 \times 10^{-2} M$ N_2O. (This work and Dainton and Logan, (46)).

3.5

KBr - SF_6 System:

The fluoride yield from solutions containing KBr $= 5 \times 10^{-3} M$ and $SF_6 = 2.4 \times 10^{-4} M$ is linear with dose up to 5 minutes of irradiation time studied using $1849 A^{\circ}$ light. From the slope (fig. 20),

$$\Phi \frac{(F^-)}{6} = 0.175 \pm 0.005.$$

At high $[SF_6]$ approximately $1.4 \times 10^{-3} M$, the yield of fluoride

Figure 20

Yield of F^- as a function of irradiation time in KBr system. $[SF_6] = 2.4 \times 10^{-4}M$, dose rate $= 3.1 \times 10^{18}$ quanta/min.

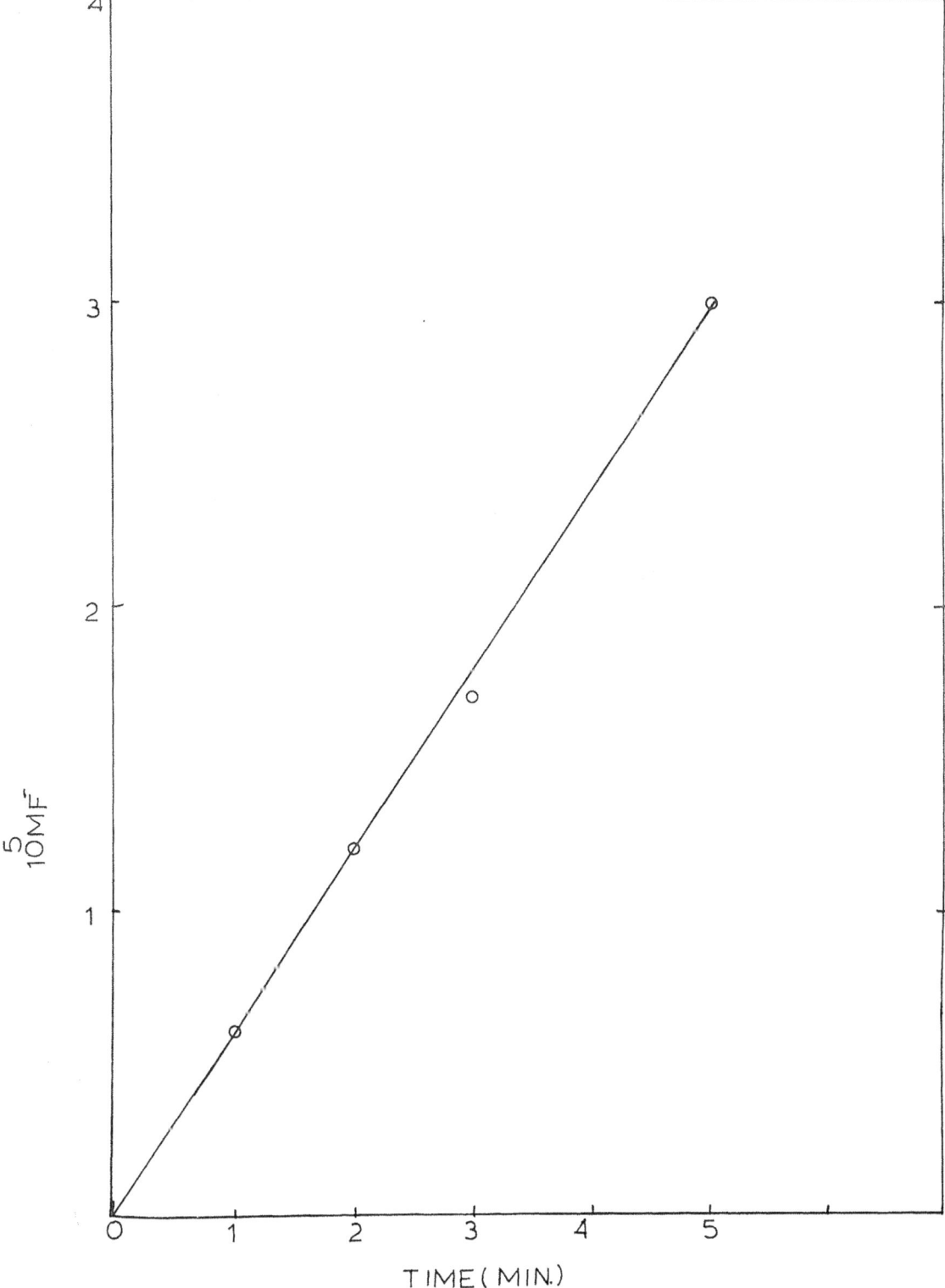

Figure 21

Yield of F⁻ as a function of irradiation time in KBr system.
$[SF_6] \simeq 1.4 \times 10^{-3}M$, dose rate $= 3.55 \times 10^{18}$ quanta/min.

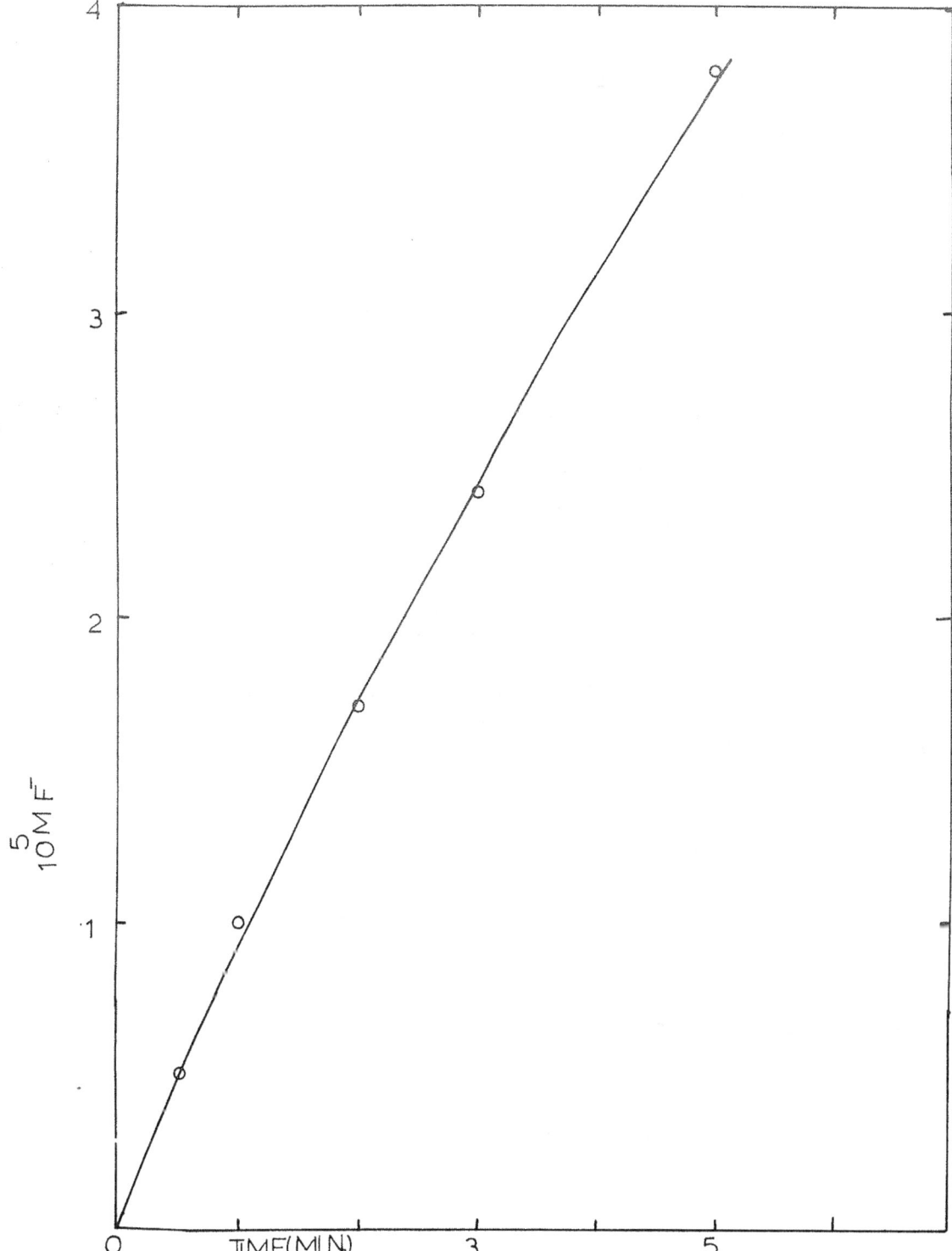

is not linear with dose and shows a continuous but slight curvature

(fig. 21). From the initial slope,

$$\Phi \frac{(F^-)}{6} = 0.33 \pm 0.02.$$

3.6

KCl - SF$_6$ System:

The fluoride yield is linear with dose up to the studied irrad-

iation time of 10 minutes for solutions in which $[SF_6] = 2.4 \times 10^{-4} M$ and

KCl = $5 \times 10^{-3} M$, using 1849A$^\circ$ light. From the slope (fig. 22),

$$\Phi \frac{(F^-)}{6} = 0.22 \pm 0.005.$$ At high concentrations of SF$_6$ approx-

imately $1.4 \times 10^{-3} M$, the yield of fluoride is not linear with dose

(fig. 23). From the initial slope,

$$\Phi \frac{(F^-)}{6} = 0.43 \pm 0.02.$$

Fig. 24 represents the variation of quantum yield of fluoride

with the scavenger concentration at the fixed time of irradiation of 1

minute. The quantum yield of fluoride shows a continuous decrease after

a small plateau with the increase of $[SF_6]$ and does not drop as in the

case of iodide system after a particular $[SF_6]$ (fig. 7). At the highest

concentration of SF$_6$ (fig. 24),

$$\Phi \frac{(F^-)}{6} = \Phi (\bar{e}_{aq}) .$$

Figure 22

Yield of F^- as a function of irradiation time in KCl system.
$[SF_6] = 2.4 \times 10^{-4}M$, dose rate $= 3.1 \times 10^{18}$ quanta/min.

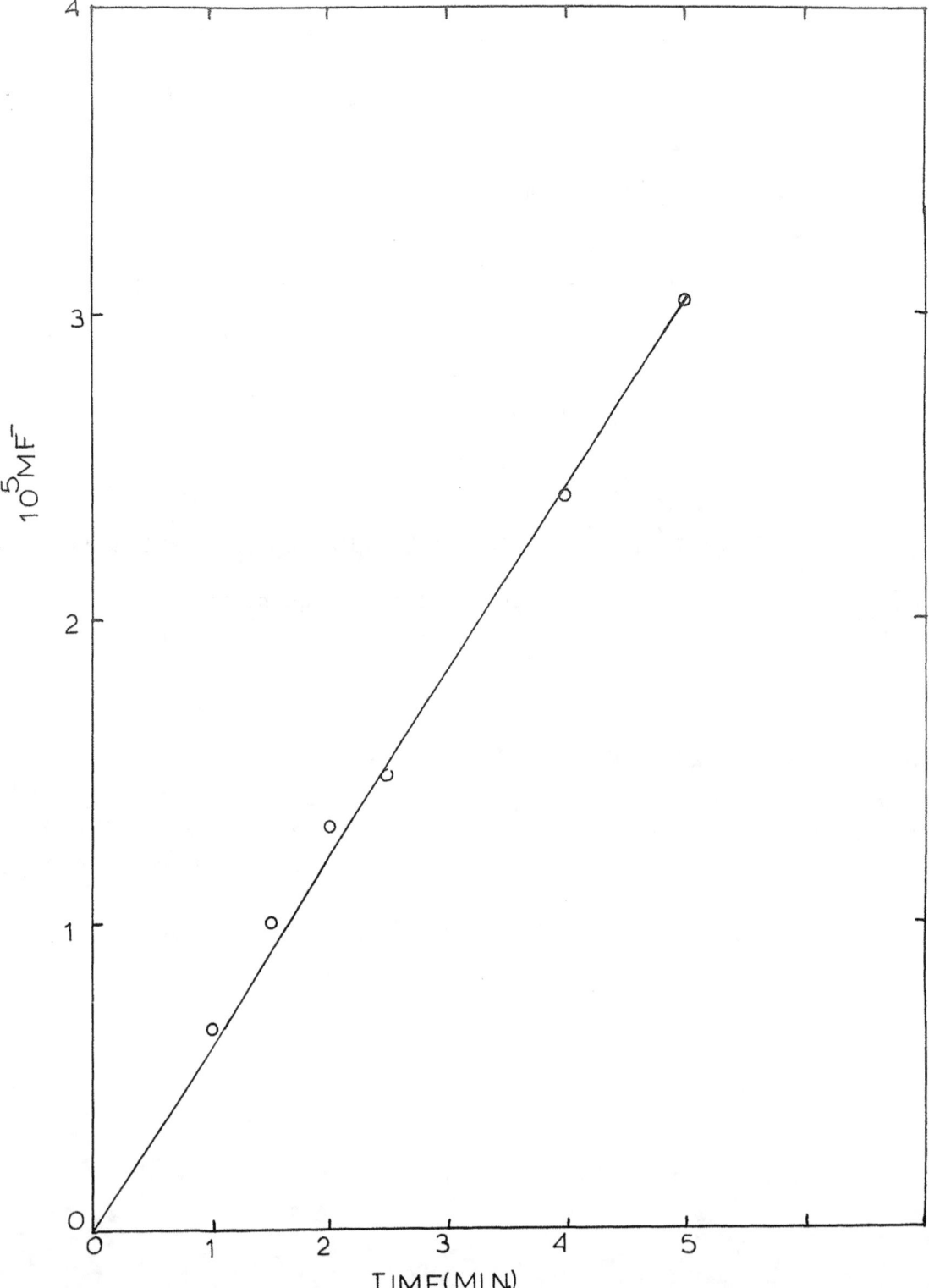

Figure 23

Yield of F^- as a function of irradiation time in KCl system. $[SF_6] \simeq 1.4 \times 10^{-4} M$, dose rate $= 3.55 \times 10^{18}$ quanta/min.

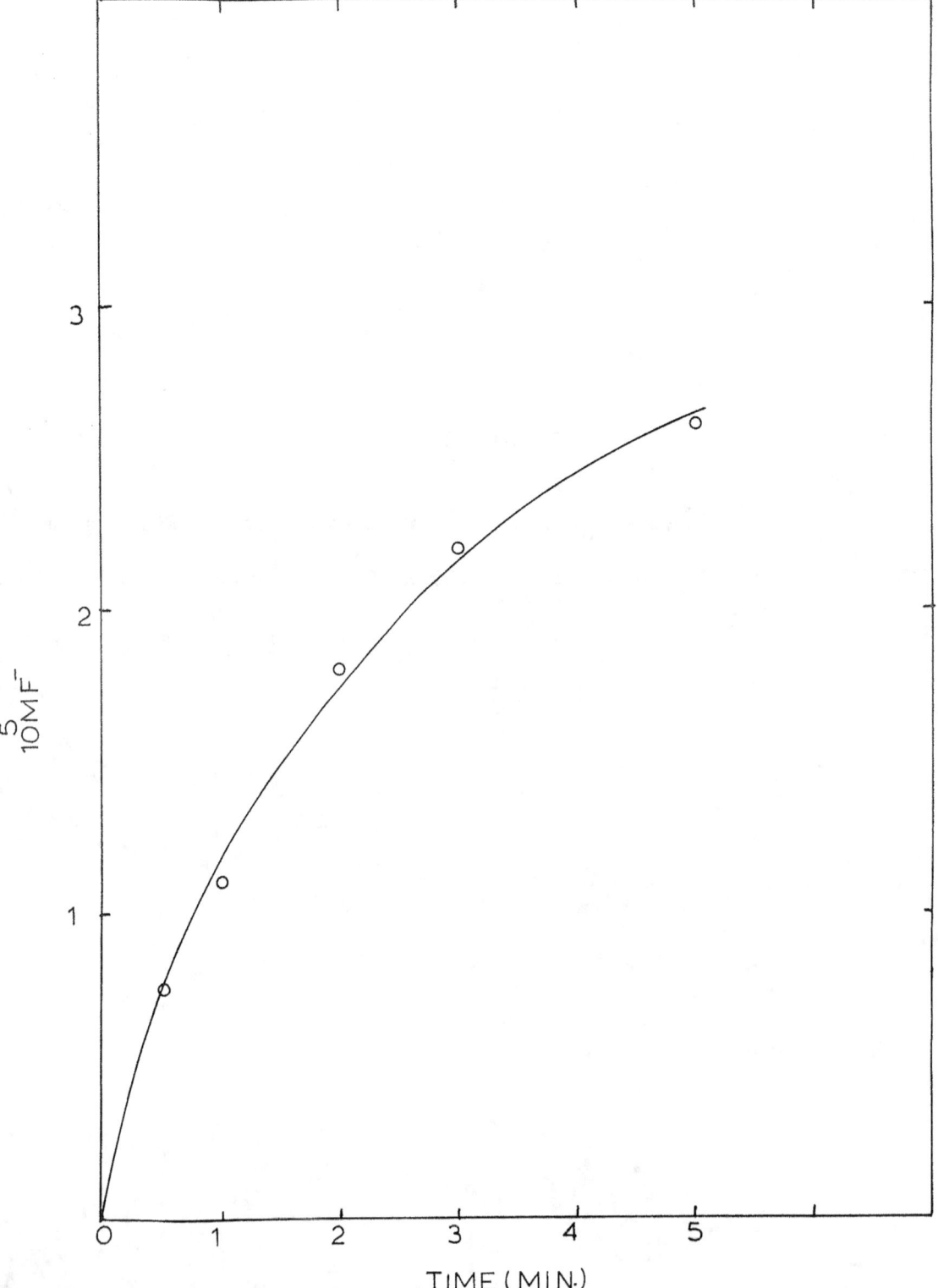

Figure 24

Quantum yield of fluoride as a function of $[SF_6]$ in KCl system. For $[SF_6] \leq 2.4 \times 10^{-4}M$, dose $= 3.1 \times 10^{18}$ quanta and $[SF_6] > 2.4 \times 10^{-4}M$, dose $= 3.55 \times 10^{18}$ quanta.

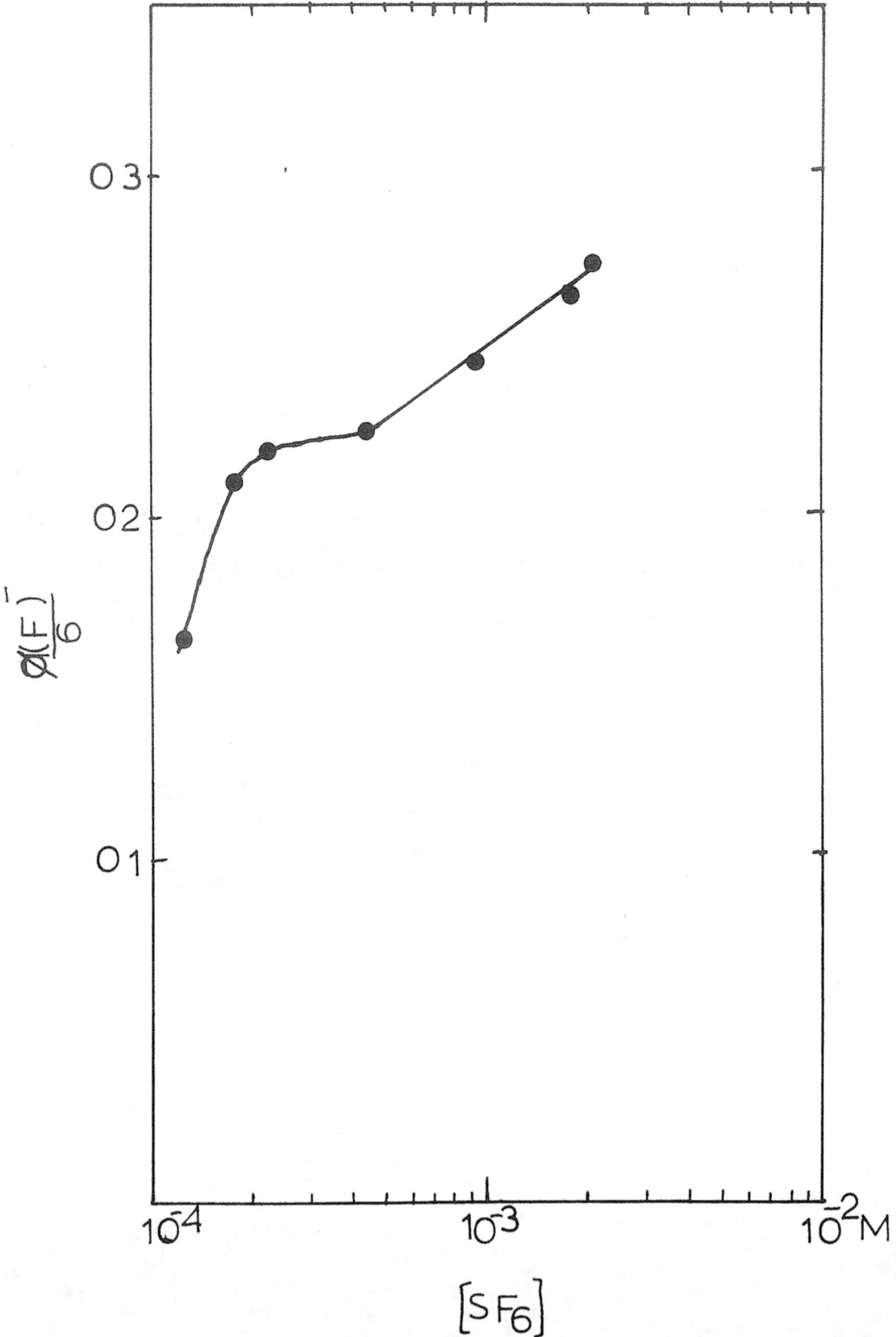

is less than

$$\phi \frac{(F^-)}{6} = 0.43 \pm 0.02$$

obtained from the initial slope from Fig. 23 .

3.7

KCNS - SF$_6$ System:

Figs. 25 and 26 represent the yield of fluoride with dose at $[SF_6]$ = 2.4 x 10^{-4}M and 1.4 x 10^{-3}M respectively, KCNS = 5 x 10^{-3}M, using 1849A^o light. The yield is linear in both cases and the quantum yields are $\phi \frac{(F^-)}{6}$ = 0.30 $\pm$ 0.005 and 0.345 $\pm$ 0.005 in the former and latter case.

3.8

KCN - SF$_6$ System:

The variation of quantum yield of fluoride with dose and its variation with time after initial measurement is shown in table 9. The quantum yields were obtained from solutions containing KCN = 5 x 10^{-3}M and $[SF_6]$ = 2.4 x 10^{-4}M, using 1849A^o light.

3.9

Fe(CN)$_6^{4-}$ - SF$_6$ System:

The variation of quantum yield of Fe(CN)$_6^{3-}$ with $[SF_6]$ obtained from solutions containing Fe(CN)$_6^{4-}$ = 5 x 10^{-3}M using 2537A^o light, is represented in fig. 27. The quantum yield shows a plateau at about 1.4 x 10^{-3}M $[SF_6]$.

KCN - SF$_6$ SYSTEM

$[SF_6]$ = 2.4 x 10^{-4} M; irradiation time = 1 min.

Time (Mts.)	$\Phi(F^-)$ (immediate)	$\Phi(F^-)$ (constant after 15 min.)
0.5	145	-
1.0	123.5	-
1.0	123.5	-
1.0	171	40
2	-	57

Table IX

Figure 25

Yield of F^- as a function of irradiation time in KCNS system.
$[SF_6] = 2.4 \times 10^{-4}M$, dose rate $= 3.1 \times 10^{18}$ quanta/min.

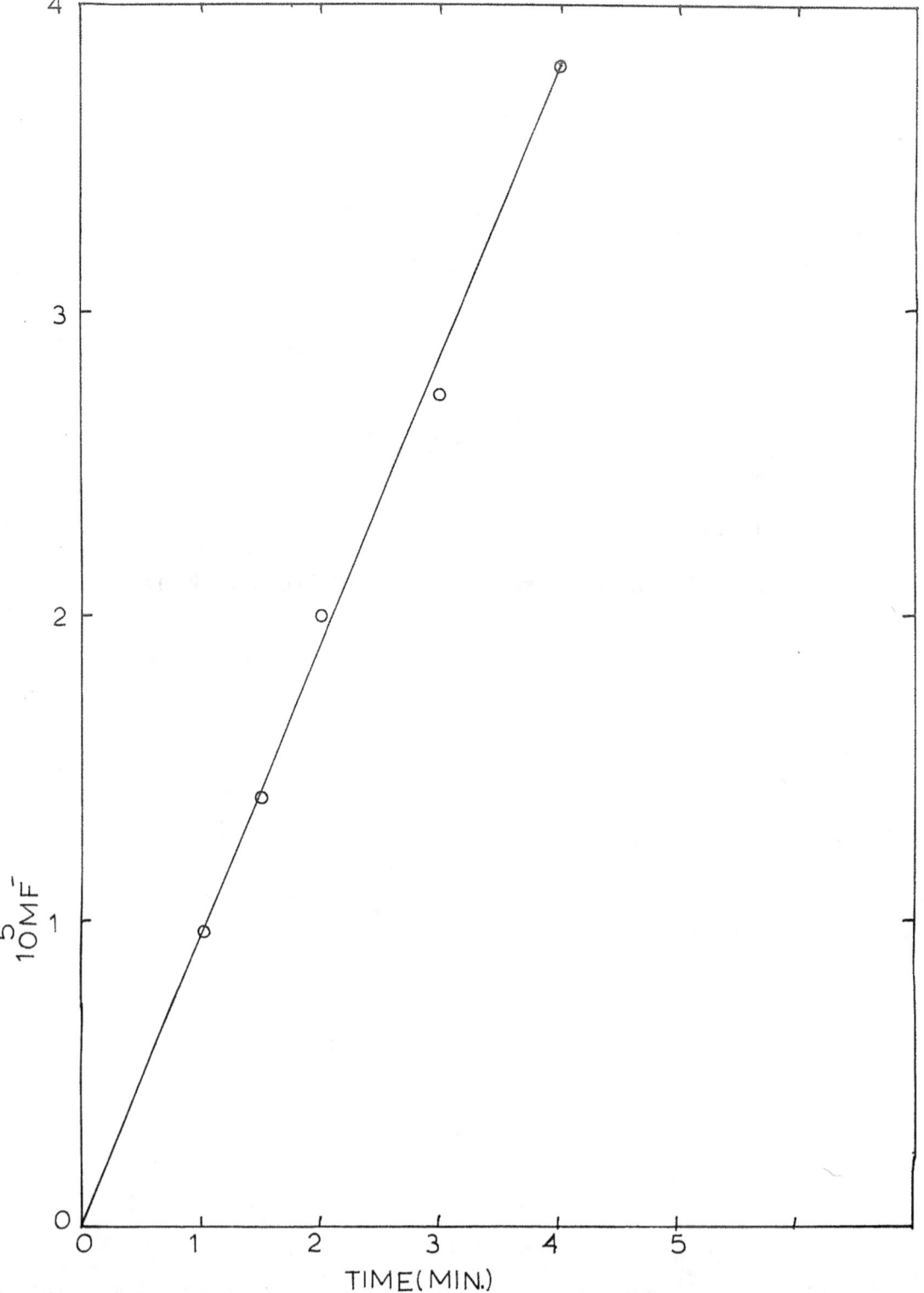

Figure 26

Yield of F^- as a function of irradiation time in KCNS system.
$[SF_6] \simeq 1.4 \times 10^{-3}M$, dose rate $= 3.55 \times 10^{18}$ quanta/min.

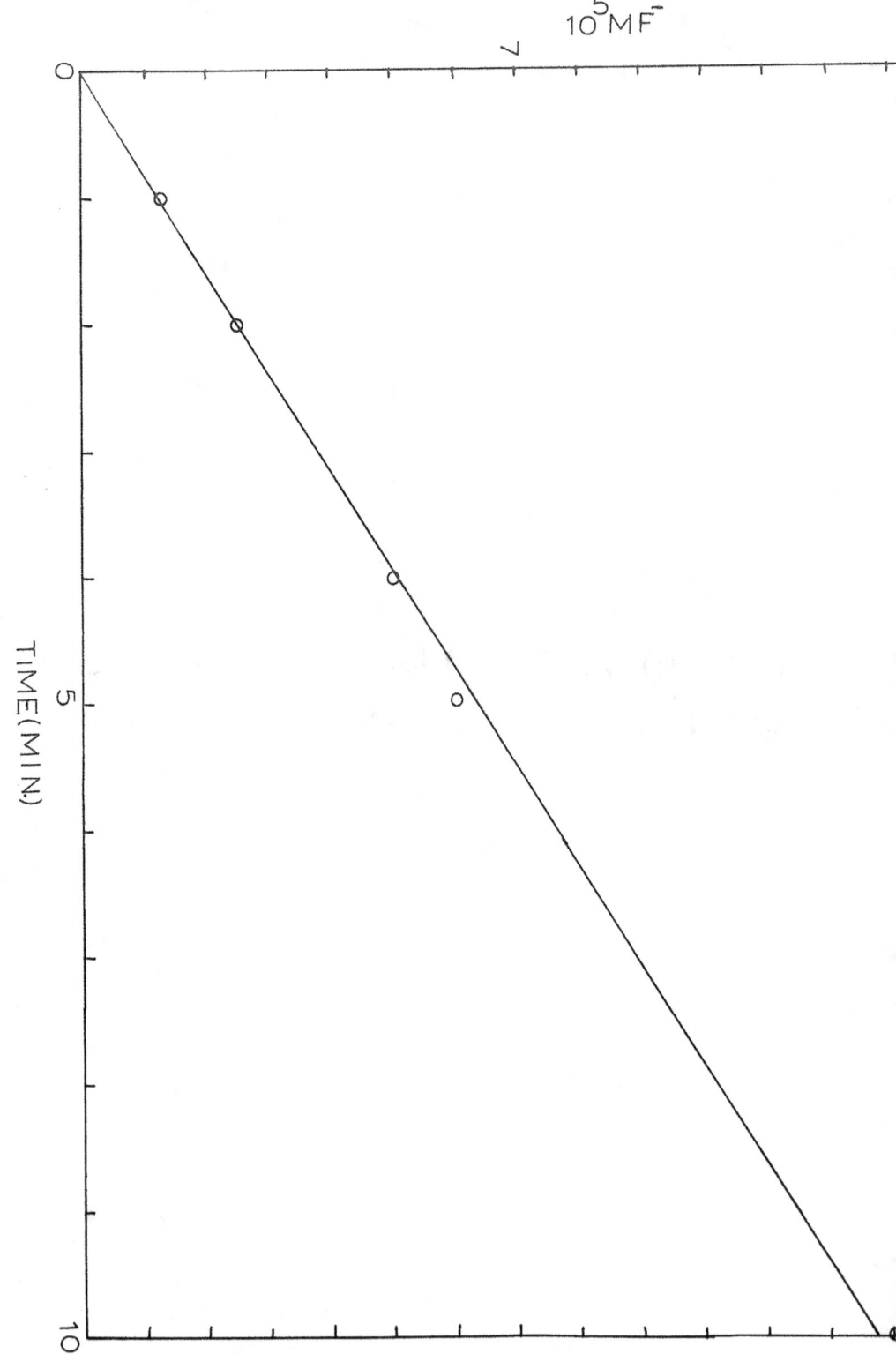

Fig. 28 shows the variation of quantum yield of fluoride with $[SF_6]$. The quantum yield again shows a plateau at about 1.4×10^{-3} M SF_6.

The yield of fluoride versus dose plot at about 1.4×10^{-3} M $[SF_6]$ (fig. 29) is linear up to irradiation time of 4 minutes and from the linear portion

$$\Phi \frac{(F^-)}{6} = 0.68 \pm 0.02 \text{ was obtained.}$$

The yield of $Fe(CN)_6^{3-}$ versus dose plot (fig. 30) shows a continuous curvature and from the initial slope,

$$\Phi \frac{(Fe(CN)_6^{3-})}{2} = 1.04 \pm 0.005 \text{ was obtained.}$$

Figure 27

Quantum yield of $Fe(CN)_6^{3-}$ as a function of $[SF_6]$. dose = 8×10^{19} quanta.

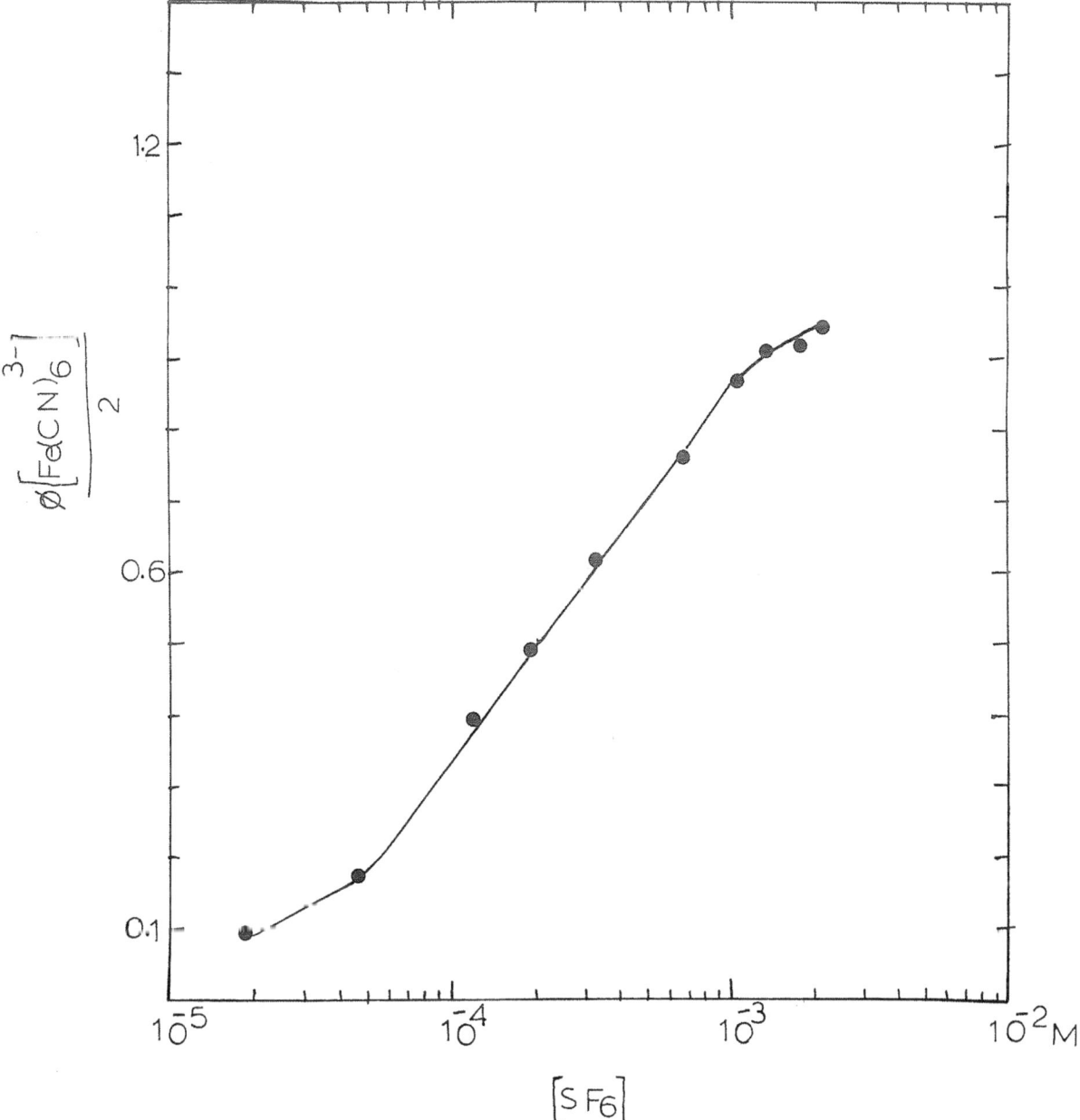

Figure 28

Quantum yield of F^- as a function of $[SF_6]$. dose $= 8 \times 10^{19}$ quanta.

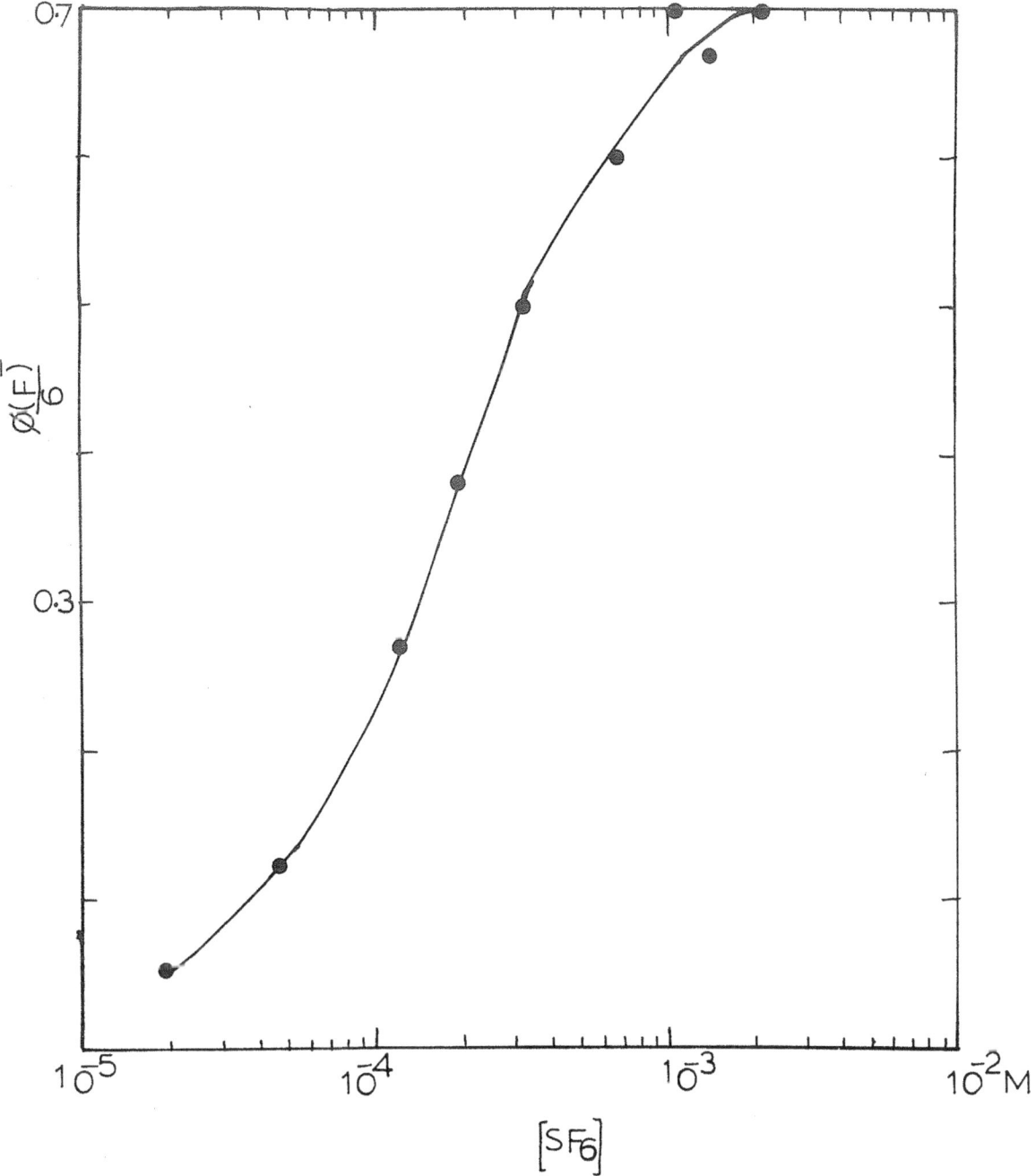

Figure 29

Yield of F^- as a function of irradiation time. $[SF_6] \simeq 1.4$ x 10^{-3}M, dose rate = 8 x 10^{19} quanta/min.

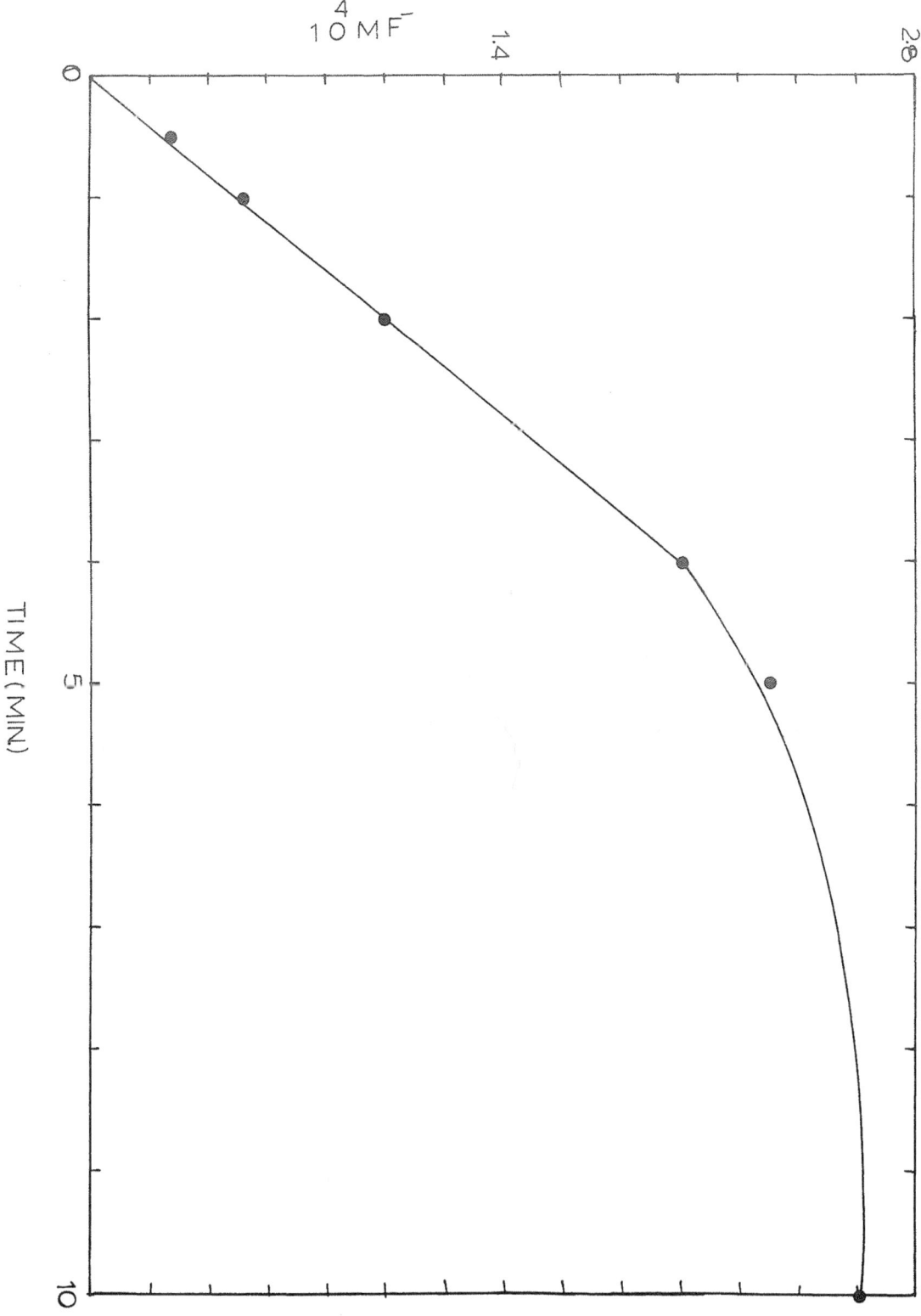

Figure 30

Yield of $Fe(CN)_6^{3-}$ as a function of irradiation time. $[SF_6] \simeq$ 1.4 x 10^{-3}M, dose rate = 8 x 10^{19} quanta/min.

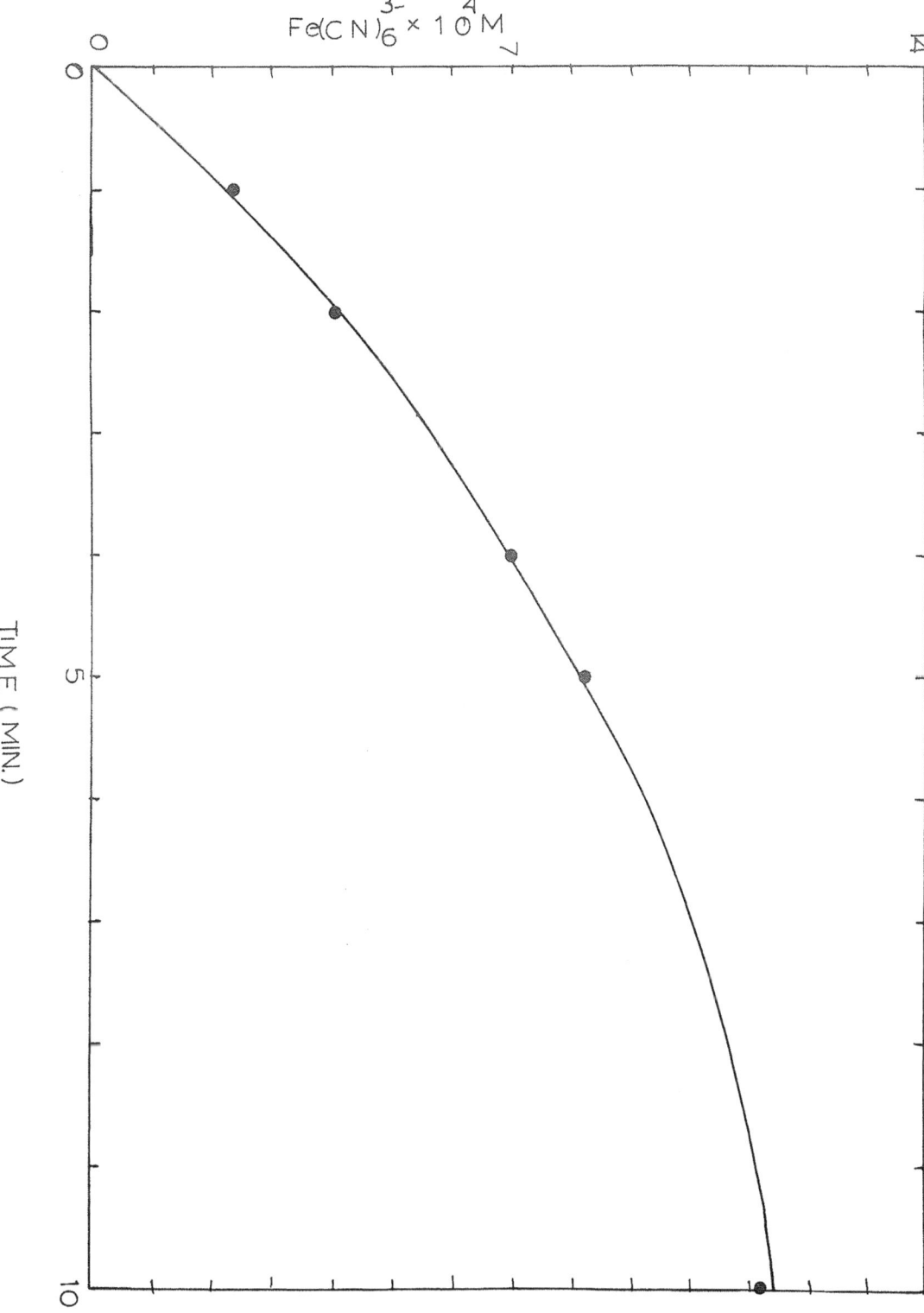

Chapter - IV

Results

Radiation - chemical

IV - RESULTS

All radiation - chemical studies were carried on the deaerated 5×10^{-3}M KI solution at its natural pH.

4.1

KI System:

Yield versus dose plot was linear with dose. The only gaseous product obtained was H_2 and $G(H_2) = 0.45 \pm 0.01$ is in agreement with literature value. The steady - state concentration of iodine was low in agreement with Allen et al (6) and $G(I_2) = 0.08 \pm 0.005$.

4.2

KI - SF_6 System:

Table 10 and fig. 31 (1) represent the variation of $G(H_2)$, $G(I_2)$ (using technique 2.9.1 a for degassing and measuring solution products after the gas analysis) and $G(F^-)$ respectively at 1.9×10^{-4}M SF_6, for which $\dfrac{G(F^-)}{6} = 2.0 \pm 0.1$ (from the initial slope of fig. 31 (1)). Table 11 and fig. 31 (2) represent the variation of $G(I_2)$, $G(H_2O_2)$ and $G(F^-)$ (technique 2.9.2 b) respectively but now $C\,\dfrac{(F^-)}{6} - 1.7 \pm 0.1$ (from the initial slope fig. 31 (II)).

Table 12 represents the variation of $G(F^-)$, $G(H_2O_2)$ and $G(I_2)$ with $[SF_6]$ using technique 2.9.2 b for measuring solution products at a fixed dose of 1.52×10^{21} ev. Since yield versus dose plots for the yield of fluoride were not carried out,

$[SF_6] = 1.9 \times 10^{-4}$ M; dose rate $= 7.6 \times 10^{20}$ ev l^{-1} min^{-1}

Time of Irradiation	$G(H_2)$	$G(I_2)$ (degassed)
1	0.59	0.185
2	0.6	0.185
3	0.56	0.21
4	0.62	0.22
5	0.61	0.28
6	0.54	0.21
8	0.6	0.28
10	0.55	0.23

Table X

$$[SF_6] = 1.9 \times 10^{-4} \, M; \text{ dose rate} = 7.6 \times 10^{20} \text{ ev l}^{-1} \text{ min}^{-1}$$

Time of irradiation	$G(H_2O_2)$	$G(I_2)$ (non-degassed)	Total $G(H_2O_2) = G(H_2O_2) + G(I_2)$
1	0.51	0.02	0.53
2	0.51	0.03	0.54
3	0.50	0.05	0.55
5	0.48	0.07	0.55
10	0.46	0.09	0.55

Table XI

$$\frac{dose = 1.52 \times 10^{21} \text{ ev}}{}$$

$[SF_6] \times 10^4 M$	$G(F^-)$	$G(H_2O_2)$	$G(I_2)$ (non-degassed)	Total $G(H_2O_2) =$ $G(H_2O_2)+G(I_2)$
0.63	3.8	0.10	0.11	0.21
1.27	5.5	0.50	0.05	0.55
1.9	7.7	0.51	0.02	0.53
2.35	8.9	0.50	0.04	0.54

Table XII

Figure 31

Yield of F^- as a function of irradiation time. $[SF_6] = 1.9$ x 10^{-4}M, dose rate = 7.6 x 10^{20} ev 1^{-1} min^{-1}. Non - degassed F^- yield (figure 31 (II)). Degassed fluoride yield (figure 31 (I)).

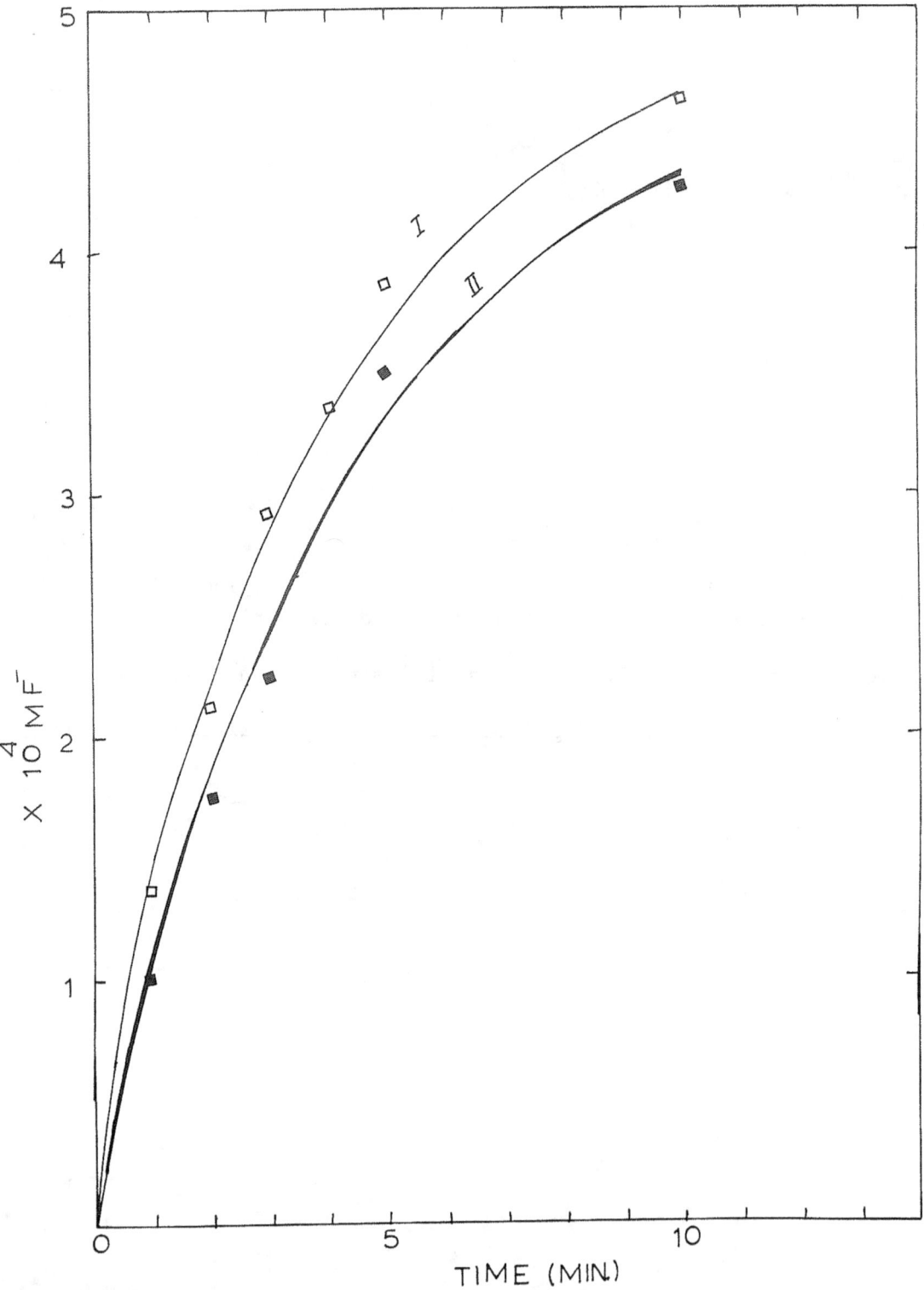

so the values of $G(F^-)$ at a fixed dose are not meaningful due to the fact that they are not linear with dose. (Fig. 31).

4.3

KI - N_2O System:

The nitrogen yield (fig. 32, table 13) increases with the increase of $\lceil N_2O \rceil$ but the increase is about 30 % more at each concentration of N_2O than the well accepted values in the literature (parentheses table 13). Table 13 also shows the variation of $G(H_2)$, $G(O_2)$, $G(H_2O_2)$ and $G(I_2)$ with nitrous oxide concentration. $G(H_2)$ is about 50 % more than the accepted literature values.

Table 14 shows the variation of non-degassed $G(H_2O_2)$ and $G(I_2)$ with $[N_2O]$. Tables 13 and 14 show that the degassed and non-degassed yields of $G(H_2O_2)$ and $G(I_2)$ are different and whereas there is a regular increase of $G(I_2)$ with $\lceil N_2O \rceil$ (table 14), there is no definite trend in the degassed $G(I_2)$ (table 13).

The yields of $G(H_2O_2)$, $G(I_2)$ and $G(O_2)$ have been shown to differ at different concentrations of KI (18, 24, 74). However, the non-degassed $G(I_2)$ from the present work is in good agreement with Hayon (24) and Anbar et al (74). Since 5×10^{-3}M KI solution is used in the present work, $G(O_2)$ and $G(I_2)$ differs with Dainton and Buxton (18) where less than 10^{-3}M, greater than 10^{-2}M solutions of KI have been used.

The degassed and non - degassed $G(I_2)$ (tables 13, 14) are not the true values since they have been measured at a fixed dose of 7.6×10^{20} ev and yield vs dose plot is linear up to a dose of 5.7×10^{20} ev

Dose $= \dfrac{100}{7.6 \times 10^{20} \text{ ev}}$

$[N_2O] \times 10^2 M$	$G(N_2)$	$G(H_2)$	$G(O_2)$	$G(H_2O_2)$	$G(I_2)$ (degassed)
0.17	2.6	0.67	0.56	–	1.22
0.23	(1.82)				
	3.28	0.67	0.79	–	1.5
0.27	(2.29)				
	3.3	0.55	0.65	–	1.37
0.29	(2.31)				
	3.27	0.63	0.67	–	1.27
0.33	(2.29)				
	3.27	0.65	0.62	–	1.58
0.34	(2.29)				
	3.59	0.61	0.78	–	1.46
0.42	(2.51)				
	3.5	0.62	0.81	–	1.77
0.49	(2.45)				
	3.4	0.61	0.75	–	1.87
	(2.38)				

Table XIII

0.57	3.5	0.64	0.80	0.41	1.72
0.86	(2.45	P.59	0.80	0.085	1.73
	3.6				
	(2.52)				
1.2	3.6	0.65	0.69	zero	2.0
	(2.52)				
1.3	3.75	0.58	0.74	zero	2.0
	(2.62)				
1.4	4.0	0.62	0.93	0.03	1.85
	(2.8)				
1.7	4.0	0.61	1.0	P.17	1.54
	(2.8)				
2.1	4.1	0.63	0.61	0.1	1.61
	(2.87)				
2.6	3.9	0.57	0.7	0.05	2.0
	(2.73)				

$[N_2O] \times 10^2 M$	$G(H_2O_2)$	$G(I_2)$ (non-degassed)
0.17	0.08	1.54
0.34	0.04	1.67
0.69	0.05	1.77
1.38	zero	1.9
2.08	0.13	1.95
2.6	0.26	1.98

Table XIV

Figure 32

$G(N_2)$ as a function of $[N_2O]$.   dose $= 7.6 \times 10^{20}$ ev.

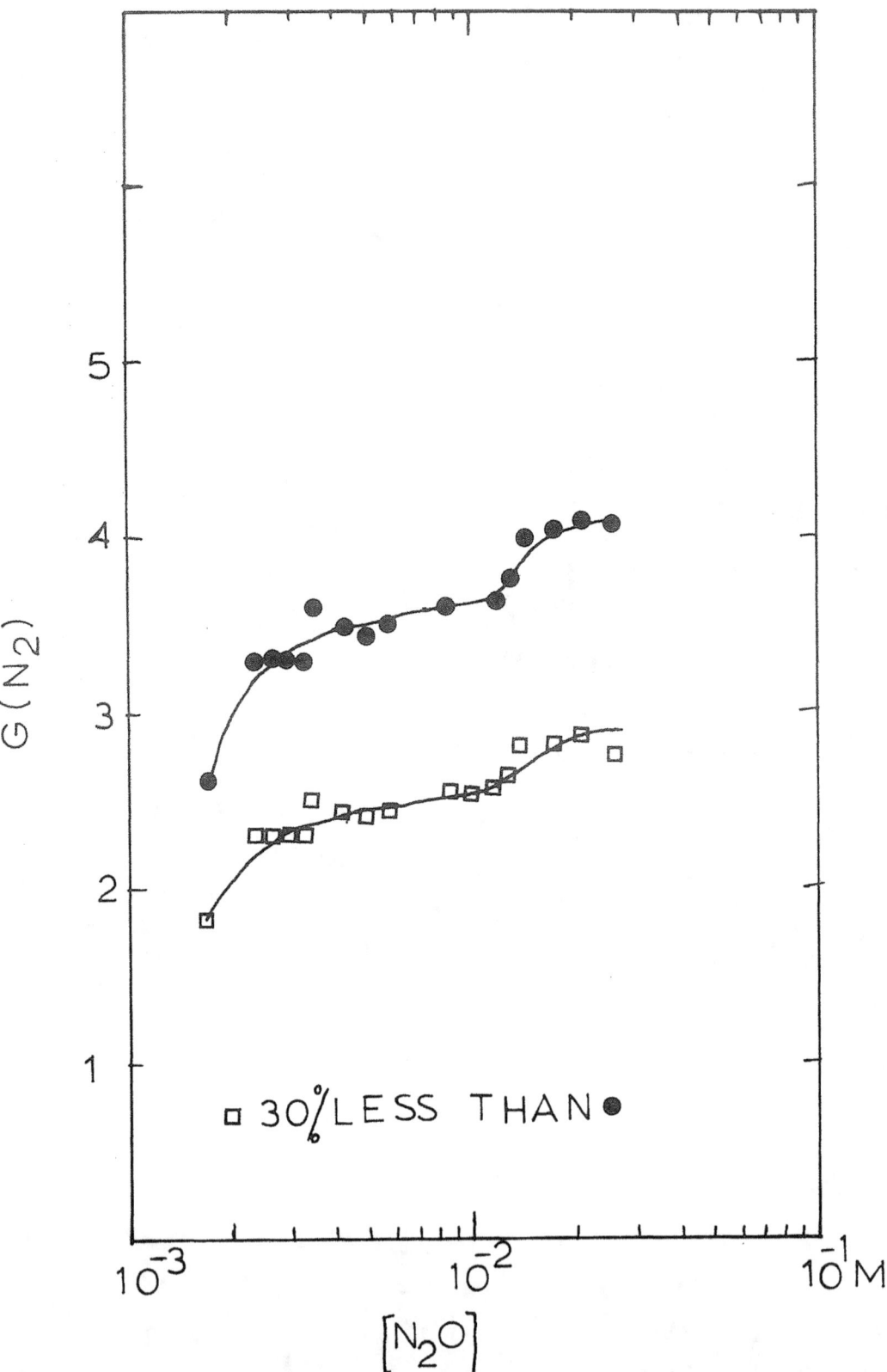

(fig. 32). Thus, $G(I_2)$ reported in tables 13 and 14 may be about 10 % less than the true values.

From the yield vs dose plot (fig. 33) at 2×10^{-2}M N_2O

$$G(N_2) = 4.1 \pm 0.1$$

$$G(O_2) = 0.63 \pm 0.01$$

$$G(H_2) = 0.65 \pm 0.02$$

and $\quad G(I_2) = 1.65 \pm 0.05$

from the linear portion of the curve for iodine yield. The gaseous products were measured by technique 2.9.1b and after that the solution products were measured.

A few experiments were done at 2×10^{-2}M N_2O and at a fixed dose of 7.6×10^{20} ev, to test the effects of degassing by different techniques, i.e. 2.9.1a, b, c.

$G(N_2) = 0.99$; $G(O_2) = 0.09$ and $G(H_2) = 0.68$ if technique a is used. The observation that there was no gaseous product except H_2 collected in the first cycle was reproducible but the yields of N_2 and O_2 collected in the second and third cycles were not reproducible; however, they are considerably less than if technique b is used for the analysis of gaseous products (fig. 32), and about 50 - 60 % of the total nitrogen and oxygen yields were collected in the first cycle using technique 2.9.1 b.

By using technique 2.9.1 c, $G(N_2) = 3.2 \pm 0.1$ and $G(O_2) = 0.33 \pm 0.01$, which is in agreement with the accepted literature values but $G(H_2) = 0.65 \pm 0.02$, which is not in agreement with the accepted

Figure 33

Yields of N_2, I_2, H_2, O_2 as a function of time. $[N_2O] =$ 2 x 10^{-2}M, dose rate = 7.6 x 10^{20} ev/1^{-1} min^{-1}.

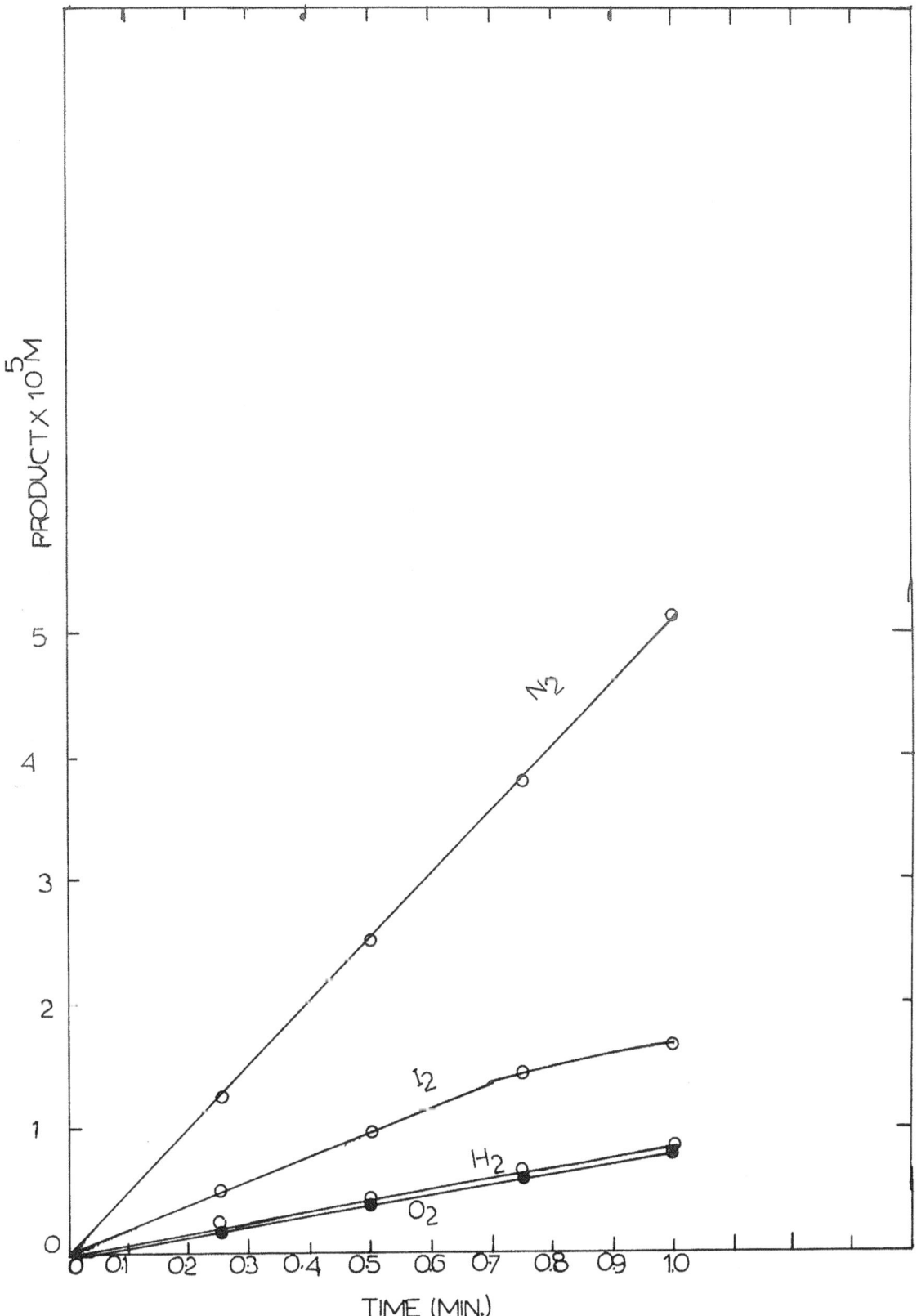

literature values (18) and is about 50 % high.

4.4

Competition studies: KI - SF$_6$ - N$_2$O Systems:

The data has been analysed in the same manner as in photo-chemical work (section 3.4), except replacing Φ by G in equations 9 - 11. The results obtained from the kinetic analysis are summarized in table 15. Technique 2.9.1 b was used for measuring gaseous products and the solution products measured after the gaseous products are shown as degassed yields, and when gaseous products were not measured and only solution products were measured, they are shown as non-degassed yields.

4.4.1

Kinetic analysis of fluoride yield:

4.4.1.1

$$[SF_6] = 1.58 \times 10^{-4} M$$

Fixed dose $= 1.52 \times 10^{21}$ ev, varying $[N_2O]$

The results of the degassed fluoride yield are shown in fig. 34 (table 15). There is much scatter of the experimental data and from the slope and intercept

$$\frac{k_{e + SF_6}}{k_{e + N_2O}} = 9 \text{ and } G(\underline{F^-})_6 = 1.66.$$

However, from the non-degassed fluoride yields shown in fig . 35,

Figure	$\dfrac{k_{29}}{k_{13}}$	$\dfrac{G(F^-)}{6}$	Figure	$\dfrac{k_{29}}{k_{13}}$	$G(N_2)$
Kinetic analysis of F$^-$ yield			Kinetic analysis of N_2 yield		
34	9	1.66	39	15 ± 1	4.25 ± 0.1
35*	3.9 ± 0.1	2.6 ± 0.1			(3.1 ± 0.1)
36(2)	5.2 ± 0.1	2.4 ± 0.1	40	8.45 ± 0.05	3.0 ± 0.1
37*	3.7 ± 0.1	2.7 ± 0.1			(2.15 ± 0.1)
36(1)	5.2 ± 0.1	2.4 ± 0.1	41	5.0 ± 0.05	3.1 ± 0.1
38*	3.8 ± 0.1	2.76			(2.25 ± 0.1)

(Figures marked with * represent the non - degassed fluoride yields and without * represent degassed fluoride yields. $G(N_2)$ given in the parenthesis has been taken from the intercepts of the lines which are displaced 30 % from the experimental lines).

Table XV

Figure 34

Degassed $[G(F^-)]^{-1}$ as a function of $[N_2O]$. $[SF_6] = 1.58$ x $10^{-4}M$, dose $= 1.52$ x 10^{21}ev.

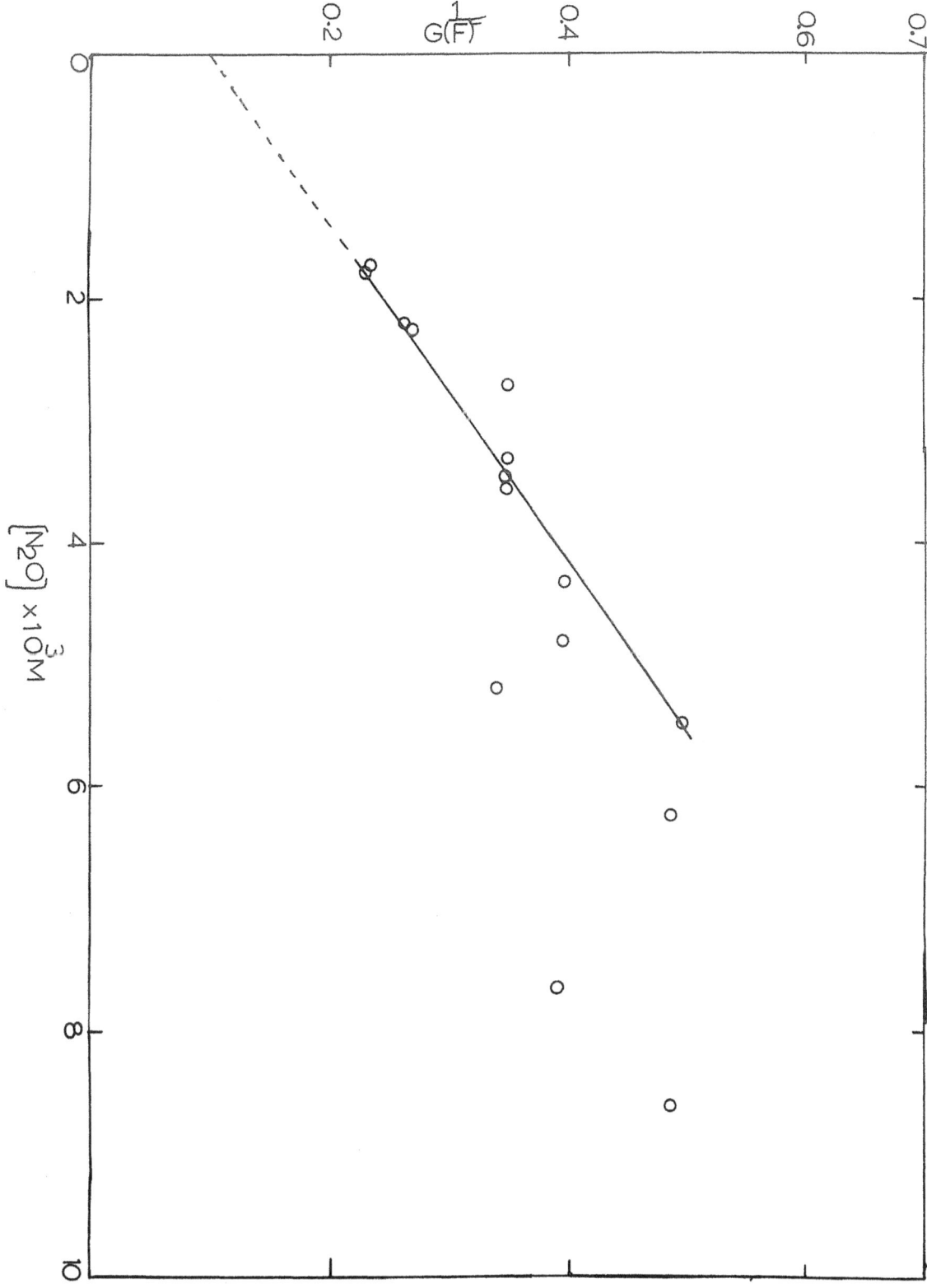

Figure 35

Non - degassed $[G(F^-)]^{-1}$ as a function of $[N_2O]$. $[SF_6] =$
1.58 x 10^{-4} M, dose = 1.52 x 10^{21} ev.

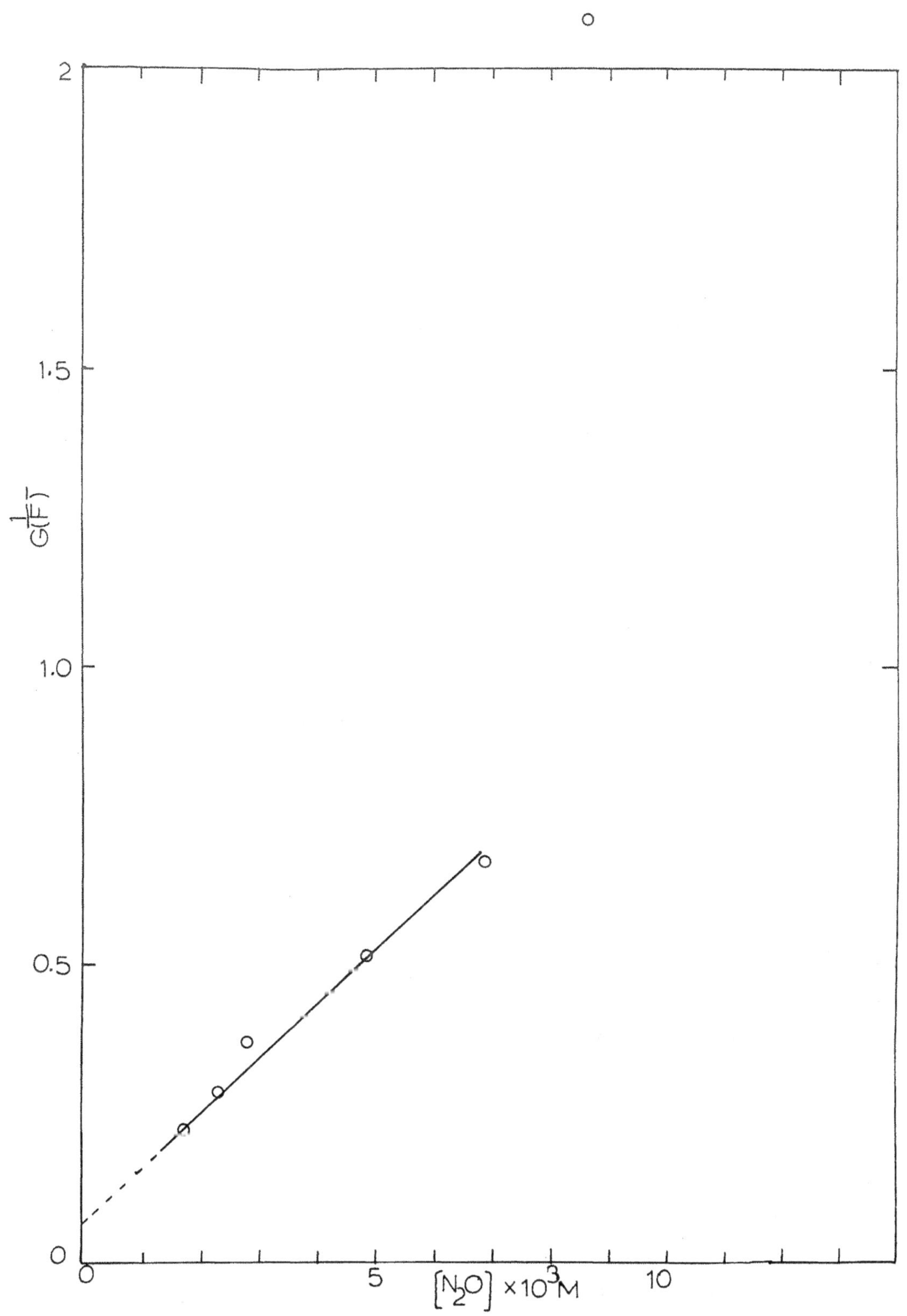

2
1.5
1.0
0.5
0
0
5
10
$[N_2O] \times 10^3$ M
$\frac{1}{G(F^-)}$

$$\frac{k_{e + SF_6}}{k_{e + N_2O}} = 3.9 \pm 0.1 \text{ and } G\frac{(F^-)}{6} = 2.6 \pm 0.1.$$

At $[N_2O] = 8.6 \times 10^{-3}M$, the value of $\frac{1}{G(F^-)}$ does not agree with the competition plot.

The degassed fluoride yields are higher than the non-degassed fluoride yield (figs. 34, 35), but at $[N_2O] = 8.6 \times 10^{-3}M$, the degassed fluoride yield is considerably greater than the non-degassed fluoride yield (table 16).

4.4.1.2

$$[N_2O] = 1.73 \times 10^{-3}M$$

Fixed dose $= 1.52 \times 10^{21}$ ev, varying $[SF_6]$

From the results of the degassed fluoride yield shown in fig. 36 (II),

$$\frac{k_{e + SF_6}}{k_{e + N_2O}} = 5.2 \pm 0.1 \text{ and } G\frac{(F^-)}{6} = 2.4 \pm 0.1.$$

Using the non-degassed fluoride yield (fig. 37),

$$\frac{k_{e + SF_6}}{k_{e + N_2O}} = 3.7 \pm 0.1 \text{ and } G\frac{(F^-)}{6} = 2.7 \pm 0.1.$$

Again the non-degassed and degassed fluoride yields are considerably different, the latter being greater at lower $[SF_6]$ and at $[SF_6] = 5 \times 10^{-5}M$. The non-degassed fluoride yield does not fit the competition

Figure 36

Degassed $[G(F^-)]^{-1}$ as a function of $[SF_6]^{-1}$. $[N_2O] = 3.46$ x 10^{-3}M, dose $= 1.52$ x 10^{21} ev (figure 36(I)). $[N_2O] = 1.73$ x 10^{-3}M, dose $= 1.52$ x 10^{21} ev. (figure 36(II)).

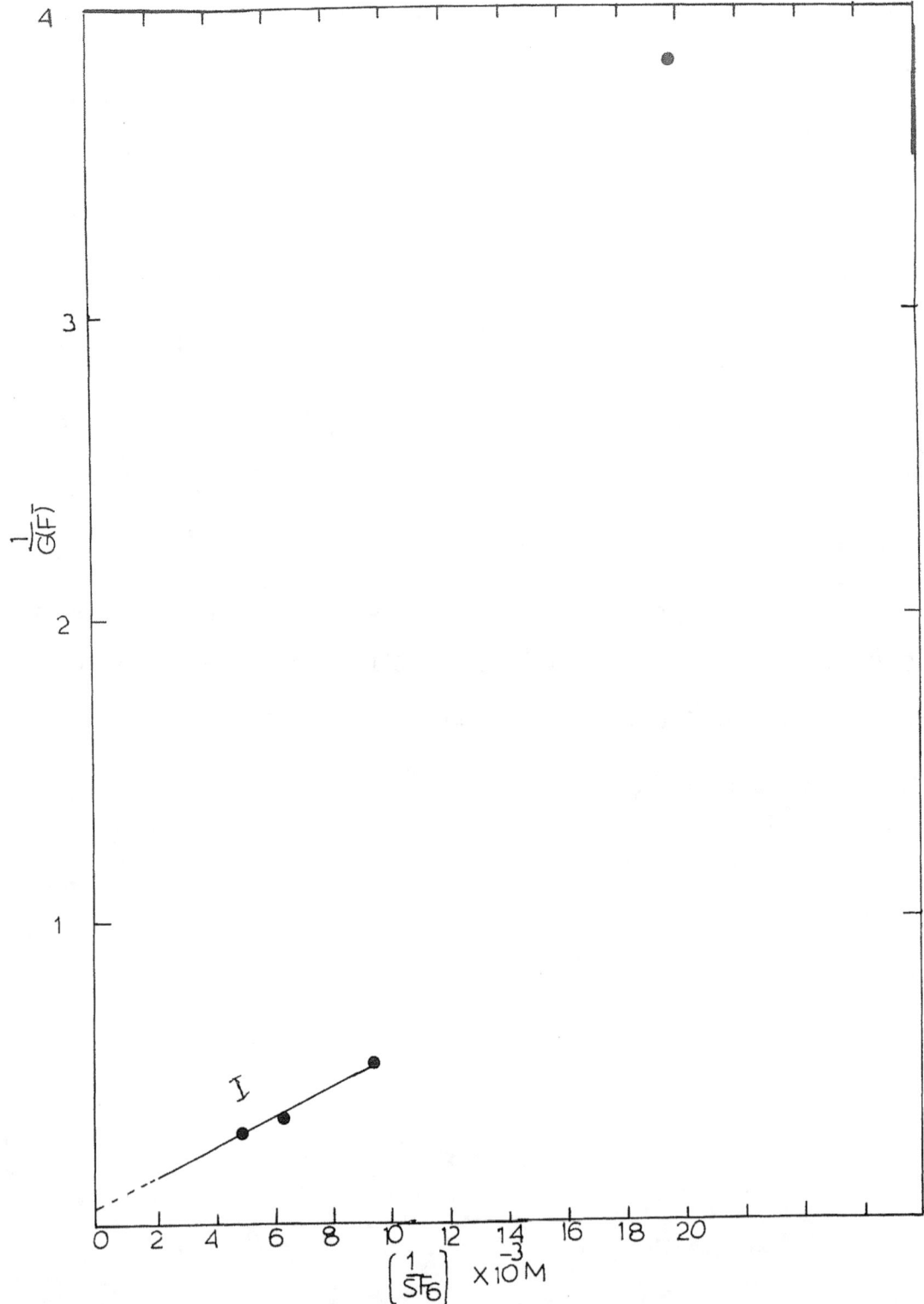

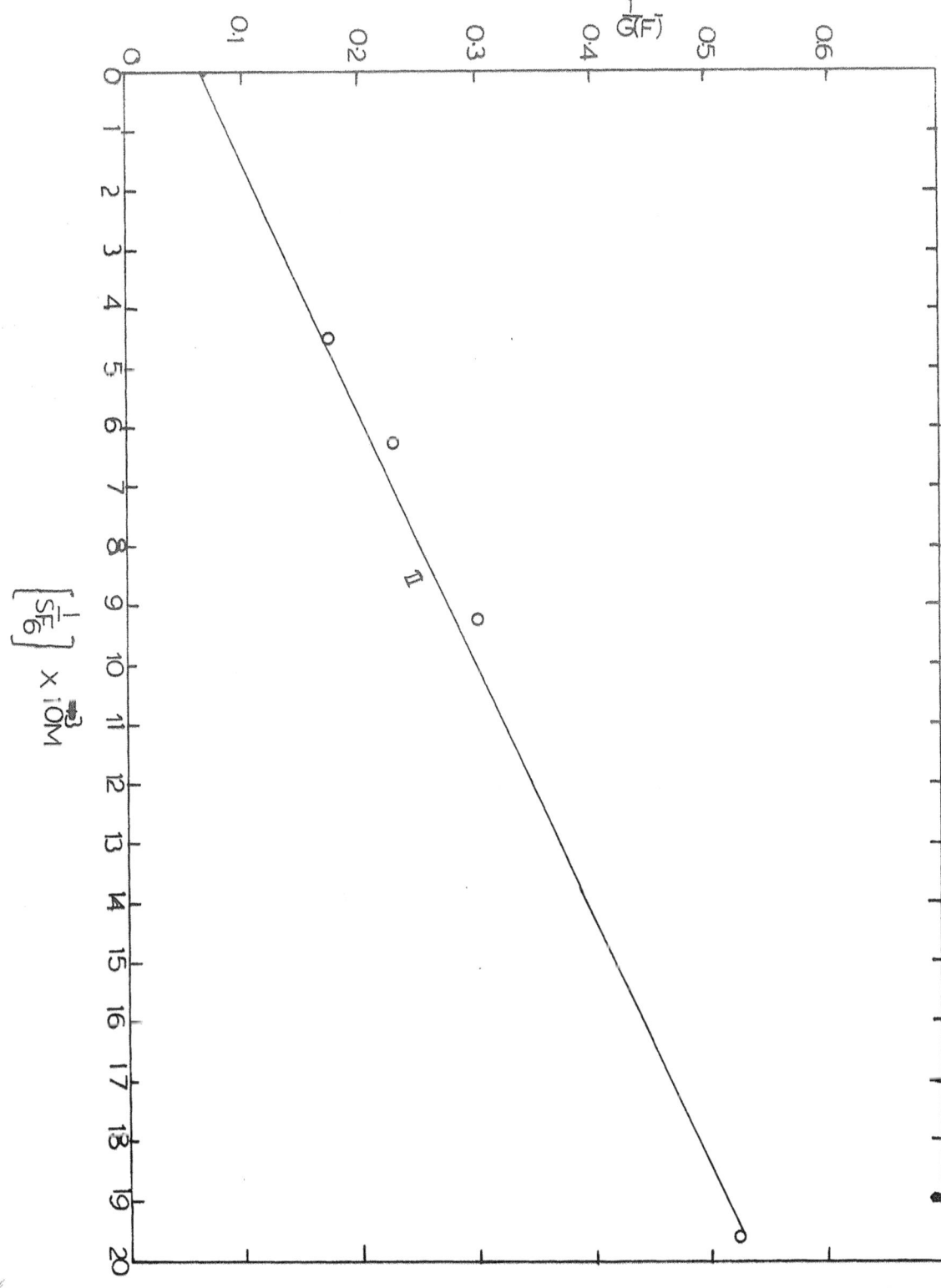

plot (fig. 37).

4.4.1.3

$$[N_2O] = 3.46 \times 10^{-3} M$$

Fixed dose = 1.52×10^{21} ev, varying $[SF_6]$.

From the degassed fluoride yields shown in figs. 36 (I),

$$\frac{k_{e + SF_6}}{k_{e + N_2O}} = 5.2 \pm 0.1 \text{ and } G\frac{(F^-)}{6} = 2.4 \pm 0.1.$$

From the non - degassed fluoride yield shown in fig. 38,

$$\frac{k_{e + SF_6}}{k_{e + N_2O}} = 3.8 \pm 0.1 \text{ and } G\frac{(F^-)}{6} = 2.76.$$

At $[SF_6] = 5 \times 10^{-5} M$, the value of $\frac{1}{G(F^-)}$ in the degassed and non - degassed cases is the same but does not agree with the competition plot (fig. 36 (I) and 38). At lower $[SF_6]$, the degassed fluoride yield is greater than the non - degassed fluoride yield.

4.4.1.4

$$[N_2O] = 8.6 \times 10^{3} M$$

dose = 1.52×10^{21} ev, varying $[SF_6]$.

The degassed and non - degassed fluoride yields are shown in table 17. The degassed fluoride yields are greater than the non - degassed fluoride yields and there is a marked difference at $[SF_6] = 1.58 \times 10^{-4} M$. No meaningful competition plot could be attempted from

$$[SF_6] = 1.58 \times 10^{-4} M, \text{ dose} = 1.52 \times 10^{21} \text{ ev}$$

$[N_2O] \times 10^3$ M	$[\frac{1}{N_2O}] \times 10^{-2}$ M	$G(N_2)$	$\frac{1}{G(N_2)}$	$G(F^-)$ degassed	$\frac{1}{G(F^-)}$	$[N_2O] \times 10^3$ M	$[\frac{1}{N_2O}] \times 10^{-2}$ M	$G(F^-)$ non-degassed	$\frac{1}{G(F^-)}$
1.71	5.83	1.88	0.54	4.26	0.234				
		(1.316)	(0.76)						
1.73	5.77	1.75	0.57	4.39	0.228	1.73	5.77	4.5	0.22
		(1.225)	(0.816)						
2.21	4.52	2.16	0.461	3.80	0.26				
		(1.51)	(0.66)						
2.25	4.43	2.06	0.485	3.72	0.268	2.28	4.3	3.5	0.28
		(1.44)	(0.69)						
2.70	3.7	2.53	0.395	2.92	0.348	2.77	3.60	2.7	0.37
		(1.77)	(0.56)						
3.3	3.04	2.5	0.4	2.9	0.345				
		(1.75)	(0.57)						
3.46	2.9	2.6	0.384	2.89	0.34	3.46	2.88	2.7	0.37
		(1.82)	(0.549)						
3.57	2.80	2.63	0.38	2.93	0.34				
		(1.84)	(0.543)						

Table XVI

Continued

4.33	2.31	2.75 (1.93)	0.36 (0.52)	2.53	0.39				
4.78	2.32	2.7 (1.89)	0.37 (0.53)	2.52	0.39	4.85	2.06	1.95	0.51
5.2	1.92	2.74 (1.92)	0.365 (0.52)	2.93	0.34				
5.48	1.82	2.8 (1.96)	0.356 (0.51)	2.02	0.49				
6.24	1.60	3.14 (2.19)	0.318 (0.455)	2.07	0.48				
7.62	1.31	3.14 (2.19)	0.318 (0.455)	2.56	0.39	6.9	1.5	1.5	0.67
8.6	1.16	3.46 (2.42)	0.289 (0.413)	2.07	0.48	8.6	1.15	0.47	2.10
						8.6	1.15	2.2*	0.46*

* Gaseous products not collected. Freeze, thaw, two cycles with dry ice + acetone bath.

$[N_2O] = 8.6 \times 10^{-3} M;$ dose $= 1.52 \times 10^{21}$ ev

$[SF_6] \times 10^4 M$	$[\frac{1}{SF_6}] \times 10^{-3} M$	$G(N_2)$	$\frac{1}{G(N_2)}$	$G(F^-)$ (degassed)	$\frac{1}{G(F^-)}$	$G(F^-)$ (non-degassed)	$\frac{1}{G(F^-)}$
0.51	19.7	3.25	0.31	0.055	18.03	0.039	25.24
1.08	9.27	3.33	0.3	0.31	3.2	0.47	2.14
1.58	6.30	3.46	0.29	2.07	0.483	0.47	2.10

Table XVII

Figure 37

Non - degassed $[G(F^-)]^{-1}$ as a function of $[SF_6]^{-1}$. $[N_2O] =$ 1.73 x 10^{-3}M, dose = 1.52 x 10^{21} ev.

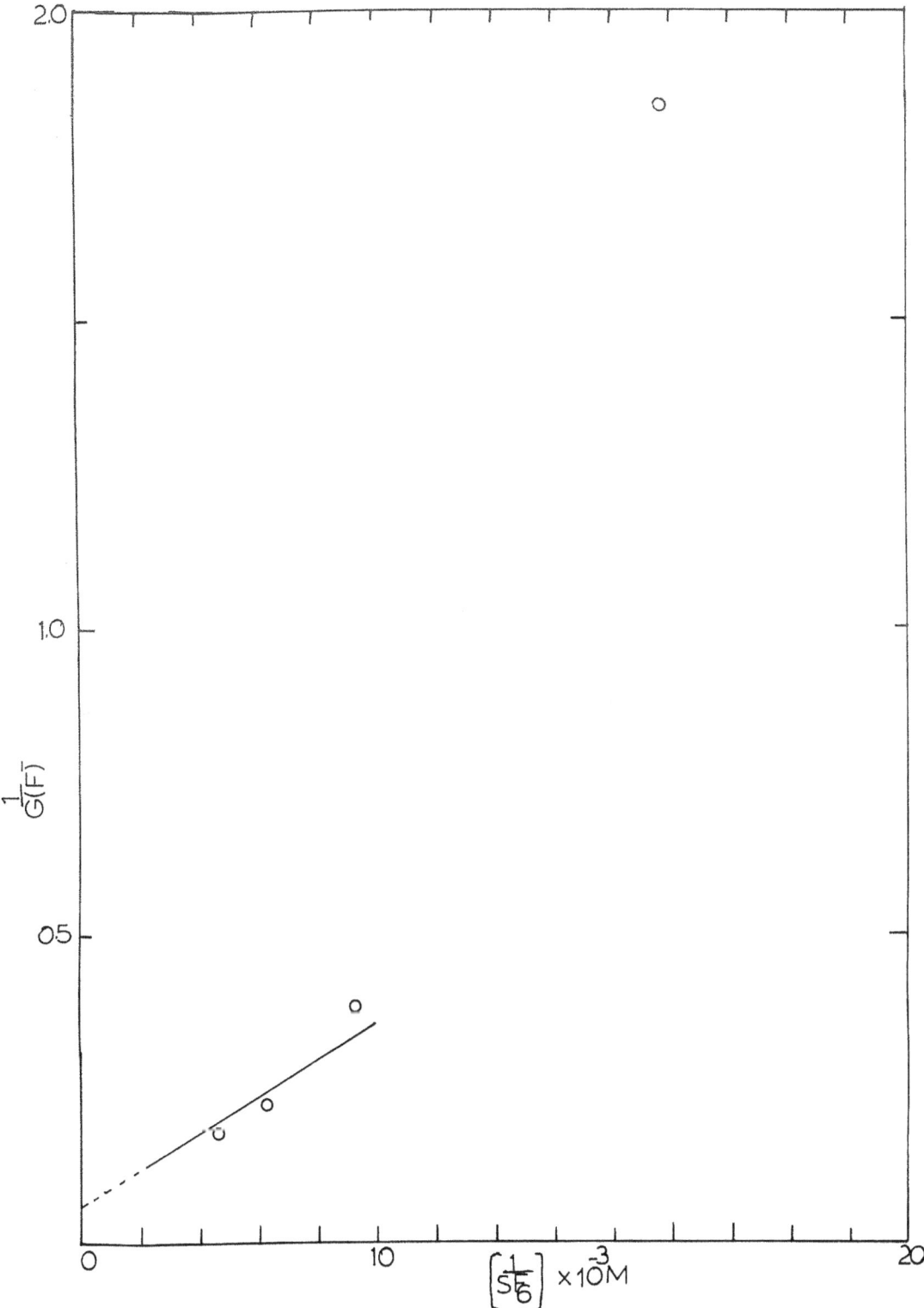

Figure 38

Non - degassed $[G(F^-)]^{-1}$ as a function of $[SF_6]^{-1}$. $[N_2O] =$ 3.46 x 10^{-3}M, dose = 1.52 x 10^{21} ev.

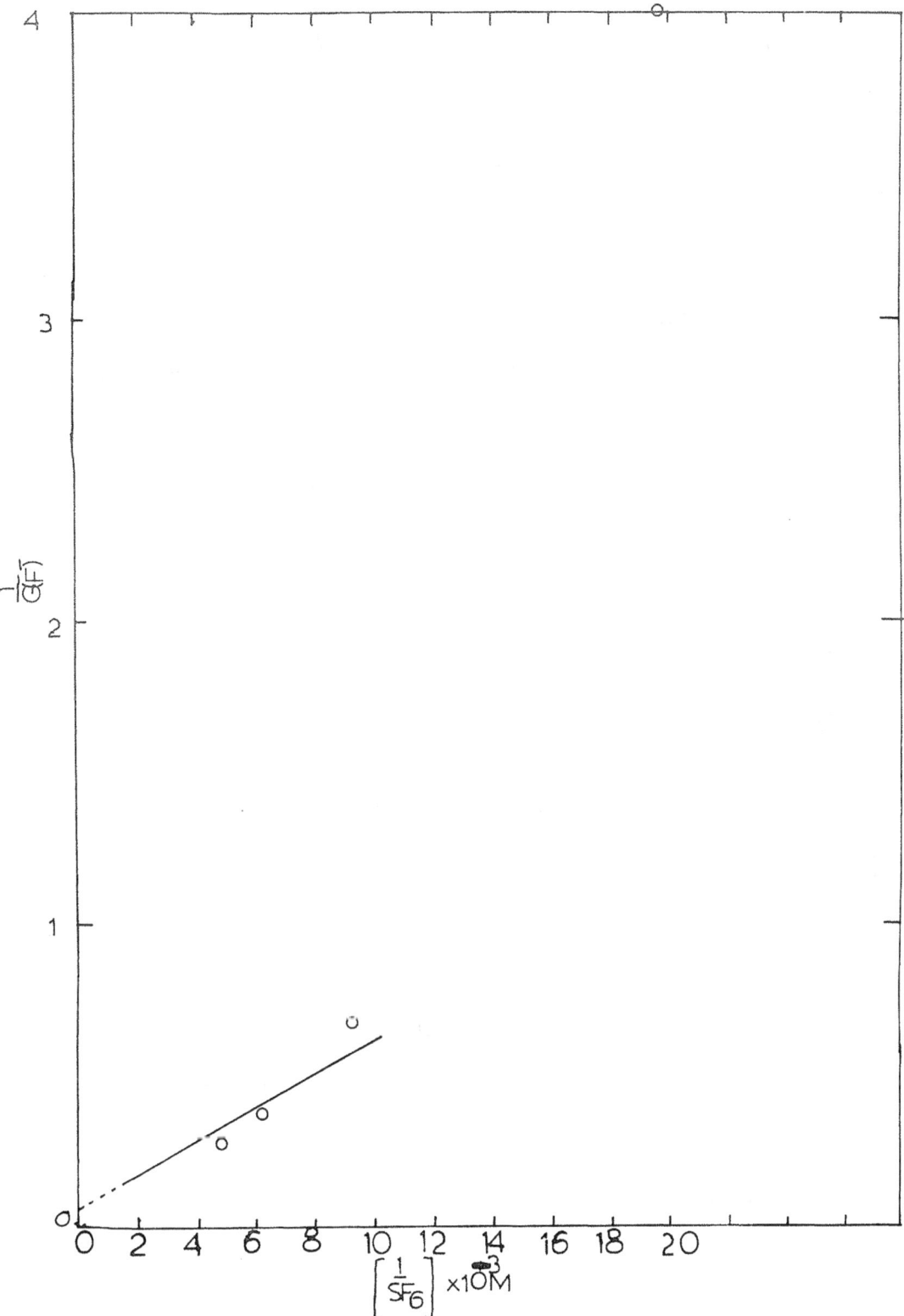

the values of fluoride yield given in table 17.

4.4.2

Kinetic analysis of nitrogen yields:

4.4.2.1

$[SF_6] = 1.58 \times 10^{-4} M$

dose $= 1.52 \times 10^{21}$ ev, varying $[N_2O]$

The reciprocals of the $G(N_2)$ versus the $(N_2O)^{-1}$ are shown in fig. 39. The experimental results are given by line II, whereas in line I the yields have been displaced by 30 % so as to give an intercept of $G(N_2) = 3.1 \pm 0.1$. In both cases

$$\frac{k_{e + SF_6}}{k_{e + N_2O}} = 15 \pm 1$$

and from the intercept in fig. 39 (II), $G(N_2) = 4.25 \pm 0.1$.

4.4.2.2

$[N_2O] = 1.73 \times 10^{-3} M$

dose $= 1.52 \times 10^{21}$ ev, varying $[SF_6]$

Fig. 40 shows the reciprocal nitrogen yield versus $[SF_6]$ plot for which line II gives the experimental results and line I the corrected results with the intercept value of $G(N_2) = 2.15 \pm 0.1$.

In both cases

$$\frac{k_{e + SF_6}}{k_{e + N_2O}} = 8.45 \pm 0.05, \text{ and from the intercept in fig. 40 (II)},$$

Figure 39

$[G(N_2)]^{-1}$ as a function of $[N_2O]^{-1}$. $[SF_6] = 1.58 \times 10^{-4}$ M, dose $= 1.52 \times 10^{21}$ ev. Line I is displaced 30 % from the experimental line II.

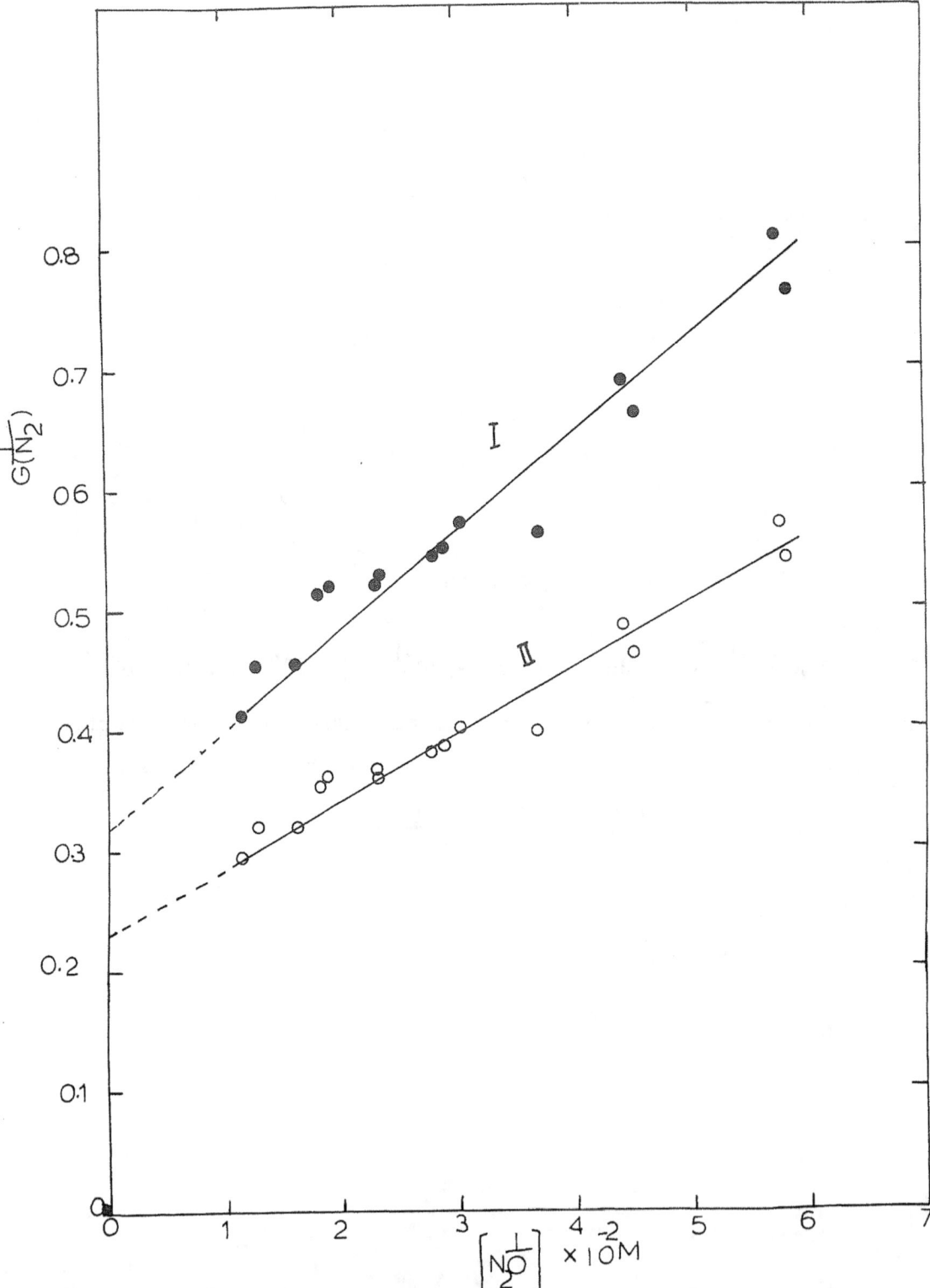

Figure 40

$[G(N_2)]^{-1}$ as a function of $[SF_6]$. $[N_2O] =$ 1.73 x 10^{-3}M, dose = 1.52 x 10^{21} ev. Line I is displaced 30 % from the experimental line II.

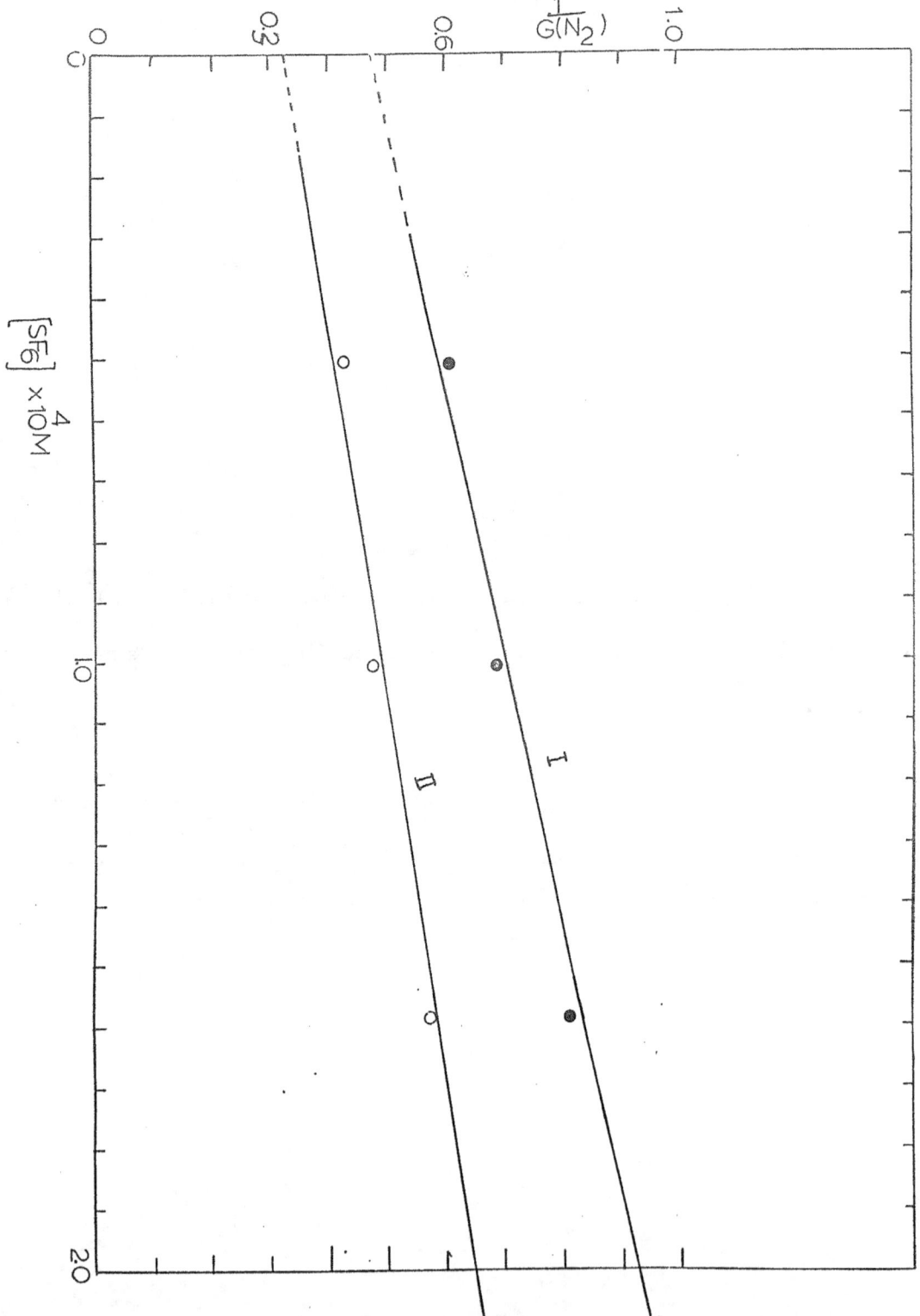

$G(N_2) = 3.0 \pm 0.01$.

4.4.2.3

$$[N_2O] = 3.46 \times 10^{-3} M$$

dose $= 1.52 \times 10^{21}$ ev, varying $[SF_6]$

Curves I and II in the reciprocal plot (fig. 41) have been obtained in the same manner as in sections 4.4.2.1 and 4.4.2.2.

In both cases

$$\frac{k_{e + SF_6}}{k_{e + N_2O}} = 5.0 \pm 0.05, \text{ and } G(N_2) = 3.1 \pm 0.1,$$

(line II); $G(N_2) = 2.25 \pm 0.1$ (line I).

4.4.2.4

$$[N_2O] = 8.6 \times 10^{-3} M$$

dose $= 1.52 \times 10^{21}$ ev, varying $[SF_6]$.

The results are presented in table 17; no meaningful competition plot could be attempted. $G(N_2)$ appears to decrease with the decrease of SF_6 concentration (table 17).

4.4.3

Kinetic analysis of iodine and oxygen yields:

No meaningful kinetic analysis of iodine and oxygen yields could be attempted. The values of the degassed iodine and oxygen yields are represented in tables 20 and 21. The degassed and non-degassed iodine yields are different and $G(O_2)$, (tables 18 and 19) is seen to decrease with the increase of SF_6 concentration and/or decrease of N_2O concen-

$$[SF_6] = 1.58 \times 10^{-4} \text{ M; dose} = 1.52 \times 10^{21} \text{ ev}$$

$[N_2O] \times 10^3$ M	$G(O_2)$	$G(H_2)$	$G(I_2)$ (degassed)
1.71	0.034	0.62	1.73
1.73	0.056	0.65	1.65
2.21	0.102	0.69	1.94
2.25	0.115	0.63	1.88
2.70	0.352	0.58	1.51
3.29	0.338	0.63	1.61
3.46	0.47	0.64	1.44
3.57	0.54	0.73	1.50
4.33	0.62	0.60	1.40
4.78	0.532	0.66	1.30
5.2	0.44	0.86	1.54
5.4	0.56	0.62	1.38
6.24	0.634	0.63	1.33
7.62	0.627	0.57	1.51
8.6	0.697	0.64	1.45

Table XVIII

$[N_2O] = 1.73 \times 10^{-3}$ M; dose $= 1.52 \times 10^{21}$ ev

$[SF_6] \times 10^4$ M	$G(H_2)$	$G(O_2)$	$G(I_2)$ (degassed)
0.51	0.61	0.385	1.24
1.08	0.61	0.188	-.76
1.58	0.65	0.056	1.65
2.22	0.59	0.037	1.29

$[N_2O] = 3.46 \times 10^{-3}$ M; dose $= 1.52 \times 10^{21}$ ev

0.51	0.63	0.66	1.54
1.08	0.62	0.62	1.44
1.58	0.64	0.47	1.44
2.06	0.68	0.31	1.63

$[N_2O] = 8.6 \times 10^{-3}$ M; dose $= 1.52 \times 10^{21}$ ev

0.51	0.58	0.61	1.61
1.08	0.62	0.68	1.53
1.58	0.64	0.69	1.45

Table XIX

Table XX

$[SF_6] = 1.58 \times 10^{-4} M$; dose $= 1.52 \times 10^{21}$ ev

$[N_2O] \times 10^3 M$	$G(I_2)$ (non-degassed)
1.73	1.42
2.28	1.63
2.77	1.58
3.46	1.43
4.85	1.52
6.9	1.54
8.6	1.71

Table XXI

$$\left[N_2O\right] = 1.73 \times 10^{-3}M; \text{ dose} = 1.52 \times 10^{21} \text{ ev}$$

$\left[SF_6\right] \times 10^4 M$	$G(I_2)$ (non-degassed)
0.51	1.33
1.08	1.50
1.58	1.42
2.22	1.43

$$\left[N_2O\right] = 3.46 \times 10^{-3}\underline{M}$$

0.51	1.54
1.08	1.50
1.58	1.43
2.06	1.78

$$\left[N_2O\right] = 8.6 \times 10^{-3}M$$

0.51	1.76
1.08	1.80
1.58	1.71

Figure 41

$[G(N_2)]^{-1}$ as a function of $[SF_6]$. $[N_2O] = 3.46 \times 10^{-3}M$, dose $= 1.52 \times 10^{21}$ ev. Line I is displaced 30 % from the experimental line II.

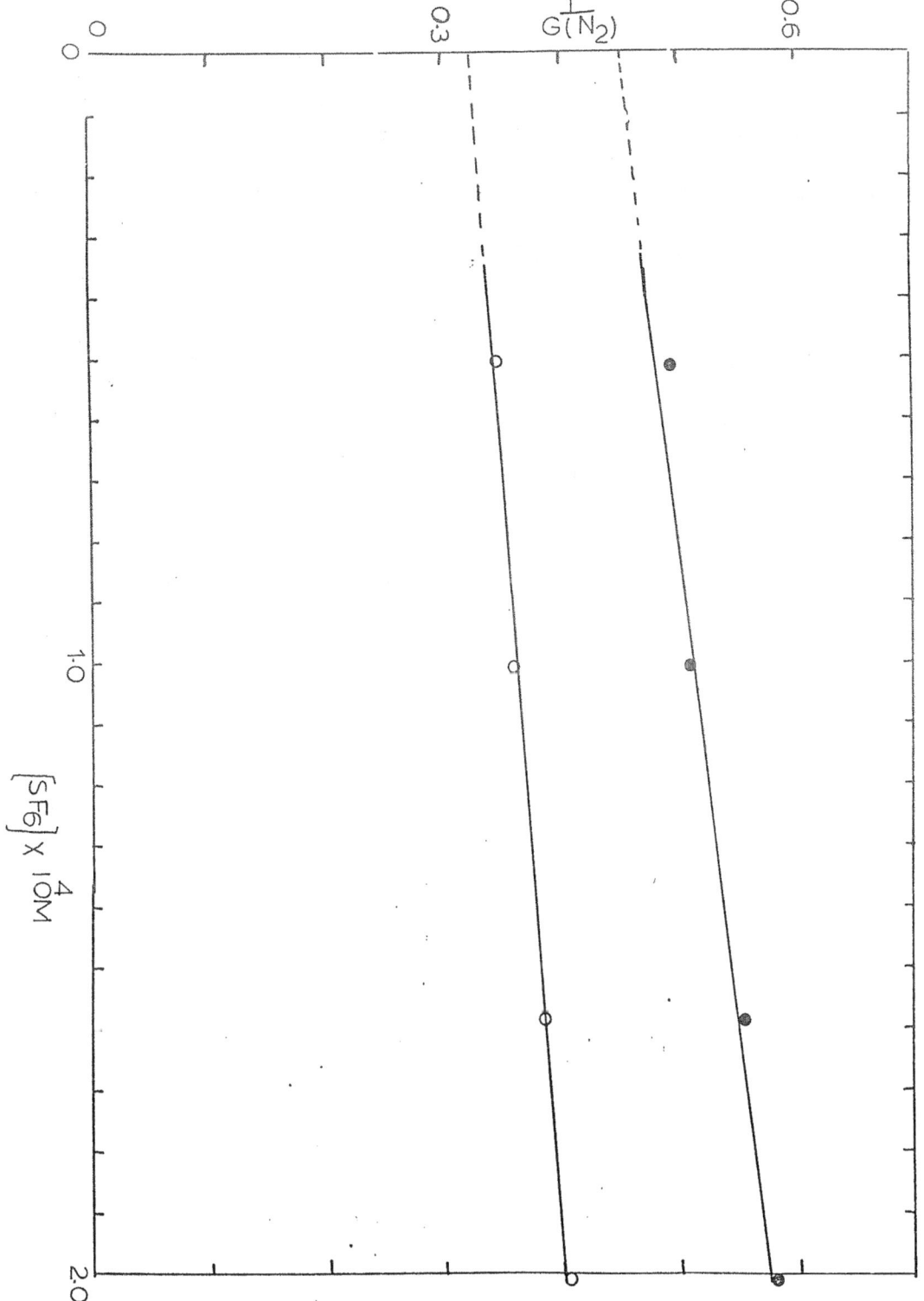

tration.

4.4.4

    Some competition studies were done at $[SF_6] = 1.58 \times 10^{-4}M$, dose $= 1.52 \times 10^{21}$ ev and varying $[N_2O]$ using technique 2.9.1 a for measuring gaseous products and after that the solution products were measured. The values of the yields of H_2, N_2, O_2, F^- and I_2 are represented in table 22. Nitrogen and oxygen yields differ significantly from those obtained by using technique 2.9.1 b (fig. 39, table 18). Also, fluoride and iodine yields are different at approximately the same scavenger concentrations (tables 16 and 18). Obviously, no meaningful kinetic analysis of the nitrogen, fluoride, iodine and oxygen yields represented in table 24 could be attempted.

$[SF_6] = 1.58 \times 10^{-4}$ M; dose $= 1.52 \times 10^{21}$ ev

$[N_2O] \times 10^3$ M	$[\frac{1}{N_2O}] \times 10^{-2}$ M	$G(H_2)$	$G(N_2)$	$\frac{1}{G(N_2)}$	$G(F^-)$	$\frac{1}{G(F^-)}$	$G(O_2)$	$G(I_2)$ (degassed)
1.52	6.55	0.54	0.09	10.9	4.9	0.20	0	1.1
1.59	6.27	0.61	0.15	6.6	5.2	0.19	0	0.95
2.55	3.92	0.61	0.16	6.25	3.3	0.3	0·	2.0
3.46	2.88	0.55	0.31	3.2	2.7	0.37	0	1.4
4.5	0.22	0.6	0.14	0.72	2.2	0.45	0	1.4
6.17	1.62	0.61	0.33	3.3	1.9	0.52	0	1.5
6.24	1.60	0.62	0	-	1.9	0.52	0	1.5
8.32	1.2	0.7	1.18	0.84	1.0	0.1	0.06	1.5

Table XXII

Motto of a Scientist

DO NOT FIT EXPLANATIONS TO THE RESULTS, BUT FIT THE RESULTS TO THE EXPLANATIONS

Chapter - V

Discussion

Photo - chemical results

V - DISCUSSION

Flash photolysis experiments by Matheson et al (76) with Cl^-, Br^-, I^-, CNS^- and $Fe(CN)_6^{4-}$ suggest that with the exception of CNS^-, photolysis of these ions leads to the formation of hydrated electrons. This conclusion is also reached from the known steady state photolysis of Cl^-, Br^-, I^- and $Fe(CN)_6^{4-}$ (46 - 50, 73, 76 - 79).

The primary photo-chemical act is assumed to be the formation of an excited state of these anions and hydrated electrons are produced from this excited state in competition with its decay to the ground state of the anion, namely in the case of I^-_{aq}:

$$I^-_{aq} \rightleftharpoons I^{-*}_{aq} \longrightarrow I^0 + e^-_{aq} \quad \dots \dots \dots \dots \dots (12)$$

N_2O has been used to a great extent in many of the photo-chemical and radiation - chemical studies as a scavenger of hydrated electrons. The absolute rate constant of the reaction of e^-_{aq} with N_2O calculated from the rate of decay of hydrated electrons in this pseudo first order reaction as measured by the pulse radiolysis technique is, $k_{13} = 8.67 \pm 0.6 \times 10^9 M^{-1} sec^{-1}$ (81)

$$e^-_{aq} + N_2O \longrightarrow N_2 + O^- \ (+H_2O \longrightarrow OH + OH^-) \ \dots \dots \dots (13)$$

Jortner et al (48) suggest the formation of an iodine atom and a solvated electron produced in close proximity by the dissociation of an excited ion. This non - homogeneous initial distribution of radicals is often referred to as the formation of radicals within a solvent cage. Various scavengers were shown (47, 48) to capture $\bar{e}_{aq}$, thus preventing the recombination:

$$I^o_{aq} + \bar{e}_{aq} \longrightarrow I^-_{aq}$$

Such a recombination of original partners is known as 'secondary (diffusive) recombination'.

The term 'primary recombination' describing recombination between radicals not yet separated by a solvent molecule, is somewhat misleading as primary recombination is the kinetic equivalent to the deactivation of the excited state. Radicals which escape secondary recombination achieve a homogeneous distribution and, if no scavenger is present, will ultimately undergo a 'bulk recombination' process.

In the absence of a scavenger, no product formation on the photolysis of aqueous iodide may presumably be due to the homogeneously distributed radicals undergoing the 'bulk recombination process'.

N_2 as the product of reaction (13) has been observed by steady-state photolysis and radiolysis techniques.

$\Phi(I_2) = \Phi(N_2)$ in the presence of N_2O (fig. 5) can be explained by the following reaction mechanism:

MECHANISM - I:

$$I^-_{aq} \longrightarrow I^o + \bar{e}_{aq} \quad \ldots \ldots \ldots \ldots \quad (12)$$

$$\bar{e}_{aq} + N_2O \longrightarrow N_2 + O^- (+H_2O \longrightarrow OH + OH^-) \quad \ldots \ldots \quad (13)$$

$$\text{OH} + \text{I}^- \longrightarrow \text{I}^0 + \text{OH}^- \ldots\ldots\ldots\ldots \quad (14)$$

$$\text{I}^0 + \text{I}^- \longrightarrow \text{I}_2^- \ldots\ldots\ldots\ldots\ldots \quad (15)$$

$$\text{I}_2^- + \text{I}_2^- \longrightarrow \text{I}_3^- + \text{I}^- \ldots\ldots\ldots\ldots \quad (16)$$

OR $\quad \text{I}_2^- + \text{I}^0 \longrightarrow \text{I}_3^- \ldots\ldots\ldots\ldots\ldots \quad (17)$

From the reaction sequence (12 - 17)

$$\Phi(\text{N}_2) = \Phi(\text{I}_2) = \Phi(\bar{e}_{aq}) \ldots\ldots\ldots\ldots\ldots \quad (18)$$

The formation of iodine due to reactions (15) and (16) is favoured over its formation due to reactions (15) and (17); however, the
stoichiometry (18) will not change even if both possibilities for its
formation are considered. Since, from the flash photolysis work of
Edgecombe and Norrish (84), Grossweiner and Matheson (82) and Hayon
et al (26), the reaction (15) and (16) is fast, I_3^- shall more likely
be formed via reactions (15) and (16). The reaction (14) is fast (83)
as measured by the rate of decay of I_2^- formed due to reaction (15) and
O^- in near neutral water should react to give the OH radical. The
increase of the yield of N_2 or iodine and thus of hydrated electrons
after about 10^{-2}M N_2O (fig. 5) similar to the increase of N_2 yield in
radiation - chemical studies, (17, 85) cannot be explained due to the
reaction of some reducing species other than hydrated electron with
N_2O, supposedly produced in C_o^{60} gamma - irradiations. In steady - state
photolysis, only hydrated electrons can be produced by the well defined
primary process, thus the increase in the yield of iodine or nitrogen

after 10^{-2} M N_2O should be due to an increase in the yield of hydrated electrons only. Freeman's non - homogeneous kinetics model (16) for the radiation - chemical yields offers an explanation of the high yields of N_2 after 10^{-2} M N_2O; however, this model may not be applicable to the photo - chemical results by virtue of the essential differences in the two primary processes. Whereas in photo - chemistry, the primary process involves the excitation of a single ion within its solvent shell, the radiation - chemical primary act is generally described in terms of the ionizations and excitations of several molecules within a much larger volume element called a spur. Hence, it is more logical to assume that the radicals produced after the primary photo - chemical act may be homogeneously distributed and 10^{-2} M N_2O is more than enough to capture all of the measured yield of hydrated electrons.

More recently, Hamill (86) has proposed that the excess yield of N_2 in radiation - chemical studies is due to the reaction of dry electron ($\bar{e}$) with N_2O and that $\bar{e}$ is also a precursor of H_2. In photo - chemistry, the electron is removed from the anion perhaps via the inter-action of the water molecules with the excited state of the anion. Indeed it is more reasonable to include these solvating water molecules as part of the excited state, thus the electron which is produced photo-chemically can only be a solvated electron. Hamill's proposal that dry electrons are the precursors for solvated electrons in radiation - chem-istry cannot be extended to the photo - chemical system in which the precursor to the solvated electron is the excited state of the anion. It has also been shown that scavengers do not interact with the excited state and thus the increase in the yield of N_2 after about 10^{-2}M N_2O,

(fig. 5) is due to the increase in the yield of hydrated electrons

only.

At low $[N_2O]$, i.e. about 10^{-4}M or lower, the low product yields

are generally interpreted in terms of the inability of the scavenger

to completely suppress the back reactions with the products e.g.

$$\bar{e}_{aq} + I_3^- \longrightarrow I_2^- + I^- \quad \dots \dots \dots \dots \quad (19)$$

$$\bar{e}_{aq} + I_2^- \longrightarrow 2I^- \quad \dots \dots \dots \dots \dots \quad (20)$$

It is therefore also important to ensure (by using low I^- concentration

or by continual stirring or by a combination of these) that I_3^- does not

accumulate in the region of the cell where the light intensity is high.

There are many anomalies in the literature due to the use of N_2O

as the scavenger of hydrated electrons which are not completely under-

stood. Thus, another scavenger of hydrated electrons, namely SF_6, which

has not been commonly used so far, should provide a useful basis for

measuring the yield of hydrated electrons.

Negative ion production in SF_6 has been studied by mass-spectroscopy

and SF_6^-, SF_5^-, F_2^- and F^- were identified from their peaks (31). The

SF_6^- and SF_5^- are the principal ions and are produced by a resonance

capture process with peak maxima at an electron energy of about 2 ev.

The resonance capture process can be described as:

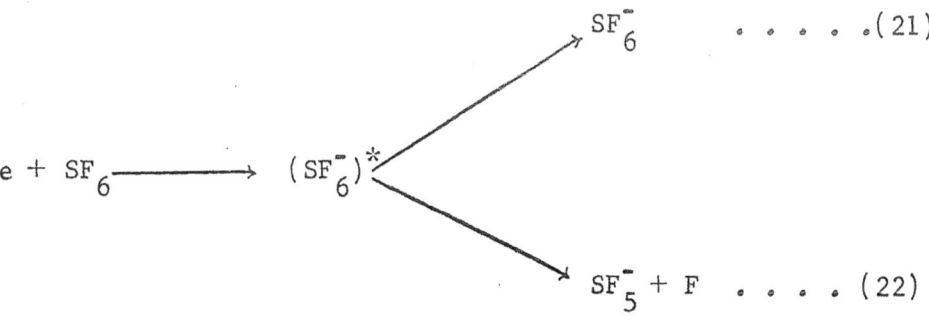

$$e + SF_6 \longrightarrow (SF_6^-)^* \begin{cases} \longrightarrow SF_6^- & \quad \ldots \ldots (21) \\ \longrightarrow SF_5^- + F & \quad \ldots \ldots (22) \end{cases}$$

The approximate equality of the peak sizes (31) indicates that (21) and (22) are of equal probability; one might expect to find similar processes leading to a more extensive dissociation, producing ions

$$SF_6 \longrightarrow (SF_6^-)^* \longrightarrow SF_4^- + 2F \text{ etc.} \ldots \ldots \ldots \ldots (23)$$

That no such reactions do in fact occur, is indicated by the non - appearance of the ions SF_4^-, SF_3^-. The dissociation may also produce a negative fluorine ion, by processes such as

$$e + SF_6 \longrightarrow (SF_6^-)^* \longrightarrow SF_5 + F^- \ldots \ldots \ldots (24)$$

$$e + SF_6 \longrightarrow (SF_6^-)^* \longrightarrow SF_4 + F + F^-, \text{ etc.} \ldots \ldots (25)$$

In a similar manner, dissociation may lead to the appearance of F_2^- ions:

$$e + SF_6 \longrightarrow (SF_6^-) \longrightarrow SF_4 + F_2^- \ldots \ldots \ldots (26)$$

$$e + SF_6 \longrightarrow (SF_6^-) \longrightarrow SF_3 + F + F_2^- \text{ etc.} \ldots \ldots (27)$$

The extrapolation of the reactions (21 - 27) to the reactions of hydrated electrons with SF_6 is rather dangerous due to the two obvious questions:

a) Is the hydrated electron similar to the electron in mass spectrometric studies?

b) Are there any specific solvent water interactions with SF_6 which will alter its gas phase characteristics?

Unlike the free electron in mass spectroscopy, the hydrated electron has an absorption spectra and thus light is absorbed by hydrated electrons obeying usual rules of spectroscopy. Hence, it is more accurate to regard the hydrated electron as some form of anion dissimilar to the electron in the gas phase.

SF_6 exhibits the lowest solubility in water at $25^{\circ}C$ of any known gas and the abnormally large entropy decrease upon its solution in water suggests a specific interaction with water, perhaps through hydrogen bonding (66).

However, believing the generally accepted cavity model for hydrated electrons that the electron formed by photolysis or radiolysis is trapped in a cavity formed by water molecules and these water molecules relax in some manner to make the trapped electron available for reaction with the solutes. The thermalized unhydrated electron ($\bar{e}$) may not react with water according to:

$$H_2O + \bar{e} \longrightarrow H_2O^- \quad \ldots \ldots \ldots \ldots (28)$$

Since the electron shells of the atoms in the water molecule are filled and therefore it is impossible for a single water molecule to accept another electron. If no major solvent perturbations on the scavenger molecule, i.e. N_2O or SF_6 are assumed, then the reactions of hydrated electrons with their scavengers are logical extrapolation of the electron reactions with their scavengers in mass spectrometer.

Thus, taking the rate of decay of hydrated electrons in the presence of SF_6 as the rate to yield the reaction products from pulse radiolysis studies (60), the reaction mechanism (II) can be proposed for the formation of fluoride from steady - state photolysis of aqueous I^- in presence of SF_6 (figs. 5 - 8):

MECHANISM II:

$$I^-_{aq} \longrightarrow I^o + \bar{e}_{aq} \quad \dots \dots \dots \dots \quad (12)$$

$$\bar{e}_{aq} + SF_6 \longrightarrow SF^-_5 + F^o \quad \dots \dots \dots \dots \quad (29)$$

$$F^o + I^- \longrightarrow I^o + F^- \quad \dots \dots \dots \dots \quad (30)$$

$$I + I^- \longrightarrow I^-_2 \quad \dots \dots \dots \dots \dots \quad (15)$$

$$I^-_2 + I^-_2 \longrightarrow I^-_3 + I^- \quad \dots \dots \dots \dots \quad (16)$$

$$SF^-_5 \longrightarrow SF_4 + F^- \quad \dots \dots \dots \dots \quad (31)$$

$$SF_4 + H_2O \longrightarrow SOF_2 + 2F^- + 2H^+ \quad \dots \dots \dots \quad (32)$$

$$SOF_2 + I_3^- + 3H_2O \longrightarrow SO_4^{2-} + 2F^- + 3I^- + 6H^+ \quad \ldots \ldots (33)$$

Asmus and Fendler (60) propose the formation of intermediates $SF_6^{\cdot-}$ and $SF_5^{\cdot}$ due to reaction (29) from the radiation – chemical studies, namely:

MECHANISM - III:

$$SF_6 + \bar{e}_{aq} \longrightarrow SF_6^{O-} \longrightarrow SF_5^{O} + F^- \quad \ldots \ldots (34)$$

$$SF_5 + 2H_2O \longrightarrow OH^{\cdot} + H_3O^+ + SF_4 + F^- \quad \ldots \ldots (35)$$

$$SF_4 + 9H_2O \longrightarrow SO_3^{2-} + 6H_3O^+ + 4F^- \quad \ldots \ldots (36)$$

$$SO_3^{2-} + (OH^O, SF_5^O, H_2O_2) \longrightarrow SO_4^{2-} \quad \ldots \ldots (37)$$

The formation of H_2O_2 is observed in radiation – chemical studies only and not in photo – chemical studies of aqueous iodide system. Accepting both mechanisms (II) and (III), the stoichiometry $\Phi \frac{(F^-)}{6} = \Phi(SO_4^{2-})$ = $\Phi (\bar{e}_{aq})$ shall remain the same. Mechanism (II) predicts zero iodine yield and the same is predictable from mechanism (III) by postulating the reaction:

$$SF_5^O + I^- \longrightarrow I^O + SF_4 + F^- \quad \ldots \ldots (38)$$

followed by reactions (15), (16), (32) and (33). We have no way of favouring any mechanism between these two since SF_6^- or SF_5^- or SF_5^0 or any other intermediate has not been observed so far from the reaction of $\bar{e}_{aq}$ with SF_6. However, mass spectrometric evidence (31) shall favour mechanism (II). The negligible amount of iodine reported from the present work may be accounted for by both mechanisms. SF_4 is known to hydrolyse very readily in aqueous media to thionylfluoride (reaction 32) which, although can be hydrolysed further to sulphite is less reactive than SF_4 (87). In the absence of any oxidant, SOF_2 has been identified mass - spectrometrically (89) and thus the most likely step is reaction (32) and not reaction (36) straight away. A standard solution of iodine is often used for the quantitative determination of sulphite and because SOF_2 belongs to the same family of compounds so its rapid oxidation by I_3^- due to reaction (33) is a fair conclusion and the fact that always some residual iodine is measured after the completion of reactions due to the photolysis of KI - SF_6 and KI - SF_6 - N_2O systems means that SOF_2 has been completely oxidised.

Nothing is known of the chemistry of SF_5^-; however, the stable coordination numbers of sulphur are two, four and six (except for crystalline sulphur) and five coordination is a notably unstable configuration. Thus it seems reasonable to assume that SF_5^- would rapidly decompose. We have chosen to represent the fate of SF_5^- by reaction (31) since SF_4 and F^- are the most direct and simplest decomposition products.

The low value of $\Phi\frac{(F^-)}{6} = 0.145 \pm 0.005$ at 2.4×10^{-4}M SF_6 (fig. 6) is generally understood in terms of solute (SF_6) depletion and disappear-

ance of $\bar{e}_{aq}$ in some back reactions but the curvature only after irradiation time = 1 minute is not understood. However, at high solute concentrations, i.e. about $1.2 \times 10^{-3}M$ SF_6, the yield of fluoride has a
smooth curvature (fig. 8) and this is definitely not due to solute depletion or the disappearance of hydrated electrons in back reactions.
Assuming homogeneous kinetics, $10^{-3}M$ SF_6 is more than enough to capture
all of the measured yield of hydrated electrons and the disappearance
of $\bar{e}_{aq}$ due to any back reactions perhaps (19) and (20) should be prevented because of the very low steady - state concentration of the reac-
tants of the back reactions. Also, the apparent fall of $\Phi(F^-)$ after
about $1.2 \times 10^{-3}M$ SF_6 (fig. 7) is not understood from mechanisms (II)
and (III).

Competition studies:

Combining mechanisms 1 and 2 or 3, equations 9 - 11 (section 3.4)
can be derived for the kinetic analysis of N_2, F^- and I_2 yields. It
should be possible to determine the total measured yield of hydrated
electrons and the rate constant ratio $\dfrac{k_{29}}{k_{13}}$ from the intercepts and
slopes of the plots obtained by these analyses. The same values of
$\dfrac{k_{29}}{k_{13}}$ and $\Phi(\bar{e}_{aq})$ must be obtained by using any of the equations 9 - 11.
Different rate constant ratio (table 1 and figs. 9 - 19) clearly sugg-
est that either the postulated reaction mechanisms are not valid or the
kinetic analyses due to these equations are wrong. Assuming the homogeneous kinetic basis of the reactions of $\bar{e}_{aq}$ with SF_6 and N_2O, these
equations are perfectly valid. Reactions (29) and (13) are pseudo first
order over the studied range of $[SF_6]$ and $[N_2O]$ and hence the rates of

these reactions should be a function of the $[\frac{SF_6}{N_2O}]$ only. Thus $\Phi\frac{(F^-)}{6}$ + $\Phi(N_2)$ should be constant and the summation should be equal to the quantum yield of hydrated electrons obtained from the intercept. The postulated mechanisms are the most reasonable that may be given to the reactions of hydrated electrons with SF_6 and N_2O. As demanded by the postulated mechanisms $\Phi(N_2)$ should equal $\Phi(I_2)$ in competition studies. It is seen from tables 2 - 7 that the summation of $\Phi\frac{(F^-)}{6}$ (degassed or non - degassed at approximately the same $[\frac{SF_6}{N_2O}]$ and $\Phi(N_2)$ is not a constant and it is less than the intercept value of $\Phi^\circ(\bar{e}_{aq})$. The summation is dependent on the concentration of the scavengers and increases with the increase of the [scavenger]. The same trend is obvious from figs. 5 and 7. The value of $\Phi^\circ(\bar{e}_{aq})$ = 0.23 $\pm$ 0.1 taken from the intercepts (figs. 9 - 19) is not the true value of the total measured yield of $\bar{e}_{aq}$ at [scavenger]$^\infty$. Since this value increases after only $\sim$ 2 x 10^{-2}M N_2O (fig. 5) and $\Phi\frac{(F^-)}{6}$ decreases after about 1.2 x 10^{-3}M SF_6 (fig. 7). $\Phi(N_2)$ is not seen to equal $\Phi(I_2)$ (tables 2 - 7) and there is a considerable difference between the non - degassed and degassed I_2 yields at about the same $[\frac{SF_6}{N_2O}]$, the latter are less than the former giving evidence to some post - irradiation reactions with iodine.

For $[\frac{N_2O}{SF_6}]$ > 10, $\Phi(I_2)$ is greater than or close to the value of $\Phi(\bar{e}_{aq})$ obtained from the intercepts of the reciprocal plots (table 8). $\Phi(N_2)$ and $\Phi(I_2)$ have unusually high values between about 3 - 4 10^{-3}M$[N_2O]$ (table 4 and fig. 18). The foregoing arguments, no formation of N_2 in the first cycle by using technique 2.9.1.a and different yields of degassed and non - degassed fluoride clearly argue against the mechanism of the reactions with hydrated electrons, and question the vital assumptions that:

a) Is the rate of disappearance of the hydrated electron in presence of SF_6 and or N_2O is also the rate to yield the reaction products?

b) Is this species paramagnetic?

c) Is the band with $\lambda_{max} \simeq 7200A^{\circ}$ due to hydrated electrons (15)?

Following alternate explanations in order to interpret the experimental data obtained from the present work are proposed:

(i) Some species, named in the literature as the hydrated electrons is not structureless but has a definite structure.

(ii) This species has structure forming characteristics.

(iii) Molecules which form structures with this species can be classed as structure makers and which break the structure of this species can be classed as structure breakers. Radicals are proposed as structure breakers only.

(iv) N_2O and SF_6 are both structure makers but SF_6 is structure breaker relative to N_2O. Henceforth the structure with N_2O shall be called as (A) and with SF_6 as (B).

The basis of these hypotheses shall be justified from the present work and the information available from the literature.

Structure of water:

Liquid water has several unique properties, indicating that there is a fundamental difference in structure between water and most other liquids. Such properties are the high melting and boiling points, the unusually high heat capacity, the decrease of the molar volume on melting and the subsequent contraction between $0°$ and $4°C$, the behaviour of aqueous solutions, etc. It was recognized very early that these anomalous properties are due to strong intermolecular interactions leading to association (57). It has generally been accepted that these interactions consist mainly of hydrogen bonding between the molecules (79). The H O molecule can participate in four hydrogen bonds, two of them
2
involving the two hydrogens of the molecule, and two the lone pairs of electrons of the oxygen and the hydrogens of two neighbouring molecules:

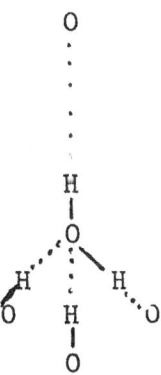

This tetrahedrally directed bonding is involved in the crystal structure of ice (90). L. Pauling (91) estimated the energy of the hydrogen bond as 5 K cal./mole on the basis of analogies with other substances. The wide range of the various estimates (95, 96, 97) stems partly from the fact that they refer to different processes; some estimates consider the transition from the hydrogen bonded ice molecule to the gas phase, some to a liquid with no dipole interactions (realised in the calculation by introducing correction factors through analogies with other substances), and some to the liquid with the dipole interactions not eliminated. Due to the large dipole moment of the water molecule, strong interactions must exist even in the non - hydrogen bonded liquid, contributing considerably to the energy needed to separate the molecules completely from each other.

Frank and Wen (92, 93) have focussed attention on the partially covalent character of the hydrogen bond; the formation of a hydrogen bond involves a contribution from delocalization energy due to electron overlap, besides purely electrostatic, i.e. dipole interaction energies, introducing some covalent character into the bond. As a consequence, large deviations from colinearity of the O - H and H·····O directions are not allowed. As Frank (92) points out, the hydrogen bond can be described in terms of the resonance scheme shown below:

I   II   III

Structure III produces a partial charge separation. As a consequence, the two molecules a and b then form additional hydrogen bonds with neighbouring molecules such as c and d in III, more easily than before the formation of the a - b hydrogen bond. Frank and Wen therefore postulate that the formation of hydrogen bonds in the liquid is a cooperative phenomenon, i.e. the bonds are not made and broken singly but several at a time, thus producing short - lived 'clusters' of highly hydrogen - bonded regions surrounded by non - hydrogen - bonded molecules. The formation and dissolution of these 'flickering clusters' is governed by local energy fluctuations. The model of 'flickering clusters' has been derived quantitatively (94) by means of a statistical thermodynamic treatment. The average cluster size ranges from 91 to 25 H_2O molecules over the temperature range from 0° to $70^{\circ}C$ with the mole fraction of non - hydrogen - bonded molecules increasing from 0.24 to 0.39 over the same range of temperature. This treatment has met with some success in accounting for the properties of water.

Frank and Wen (93) suggest a lifetime for the flickering clusters possibly as short as 10^{-10}, or 10^{-11} sec. based on relaxation times. The clusters form and melt as a consequence of local energy fluctuations. The clusters are considered to be imbedded in and in equilibrium with monomeric 'unbonded water.' In the latter, the hydrogen bonds are broken but each molecule is participating in strong dipole - dipole interactions with neighbouring molecules. The clusters grow by attachment of unbonded molecules and melt to give the latter. At a given temperature the total number of hydrogen bonds existing in the liquid and the equilibrium between clusters and unbonded water are determined by the requirement that the free energy of the system should be a minimum.

Structure of SF_6:

In sulphur the mean distance of the 3d electron from the nucleus is far greater than the mean distance of a 3s or 3p electron (98, 99) which implies that its inclusion will not significantly increase the overlapping power of the hybrid (d^2sp^3 for SF_6). A satisfactory increase could be obtained if for some reason, the d orbitals were contracted. This has been shown by theoretical calculations in the case of SF_6 (99). Calculations for SF_6 suggest that the 3d orbital contracts in such a way that its maximum occurs at approximately the S - F distance. An electron in the sulphur 3d orbital is then closer to the fluorines than to the sulphur nucleus. The usual concept of ionic character is scarcely applicable. However, since the outer atoms are electronegative, it seems probable that there is a large amount of pure ionic character in the bonds, the region of overlapping charge is much nearer to the more electro - negative centre as common sense requires such as would be represented in the extreme case by S_6^+ (F_6^-), and in less extreme ways by a series of valence - bond structures with the negative charge being found equally on all six fluorines such as:

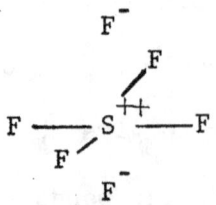

If a covalent bond model for SF$_6$ is to be retained then the d - orbitals are apparently much too weakly bound and diffuse to participate appreciably in bonding. Thus the weakly bound d - orbitals on sulphur atom would be strongly polarized by electronegative ligands through the partial withdrawal of charge from the weakly bound orbitals of the central atom and a tightening of the remaining charge to a size compatible with binding, and the suggestion that SF$_6$ is held by d^{2}sp^3 covalent bonds using sulphur 3d orbitals is untenable.

Structure of N$_2$O:

L. Pauling (100) has described the structure of N$_2$O in the following manner:

This molecule is linear, with the oxygen atom at one end. It contains 16 valence electrons; and it is seen that these can be assigned to the stable L orbitals of the atoms in the following reasonable ways:

A $:\ddot{\text{N}}{}^{\cdot} = \text{N}^+ = \ddot{\text{O}}:$

B $:\bar{\text{N}} = \text{N}^+ = \ddot{\text{O}}$

C $:\text{N} \equiv \text{N}^+ - \ddot{\text{O}}:^-$

Each of these three structures involves four covalent bonds and a separation of charge to adjacent atoms. (structures A and B differ in that in A the double bond between N and N is formed with use of p$_z$ orbitals and that between N and O with p$_y$ orbitals, and in B they are reversed.

Other structures that might be written are recognized at once as being much less stable than these, such as

$$D \qquad :\ddot{N}^- \!\!-\! N^+ \equiv \overset{+}{\underset{\cdot}{O}}$$

on which instability is conferred by the arrangement of electric charges, and

$$E \qquad :\ddot{N}^- = \ddot{N} - \overset{+}{\ddot{O}}:$$

$$F \qquad :\ddot{N} - \ddot{N} = \ddot{O}:$$

with instability arising from the smaller number of covalent bonds. We accordingly expect the normal state of the molecule to correspond to resonance among structures A, B and C. The molecule's interatomic distances and force constants are not those corresponding to any one structure alone, but to resonance among them. Its electric dipole moment is not large, but is close to zero, the opposed moments of the structures cancelling each other; the experimental value from the microwave spectrum (101) is 0.166 ± 0.002D, with the direction not known.

Faulty basis of the kinetic reactions of hydrated electrons:

The hydrated electron is commonly understood as the thermal dry electron self trapped in a cavity formed by water molecules. Theoretically speaking, since the electron shells of the water molecule are filled, the extra electron can not be accomodated in it. Naturally a single electron in a cavity shall have unpaired spin and the species should be paramagnetic. Though the limits of experimentation have prevented from seeing this paramagnetic species in liquid water by e.s.r. (electron spin resonance) technique, yet numerous studies have been carried out by glassy matrix isolation at liquid nitrogen temperature. This species has been observed only in alkaline glasses and not in neutral glasses by e.s.r. The singlet e.s.r. band (102 - 104) is generally understood as the trapped electron which becomes mobile on warming the matrix. The trapped electron in aqueous alkaline glasses is perhaps interpreted as the proof of the existence of the hydrated electron. Another proof often cited in support of the existence of hydrated electrons and their kinetic reactions is due to the Brønsted - Bjerrum relationship plot in that the hydrated electron gives a unit negative charge (38, 39) Czapski and Schwarz have assumed a mechanism based on the reactions of H_2O^- which may not be true (38) and Dainton et al (39) obtain the unit negative charge from the slope of the plot which has a large experimental error. The proof that the hydrated electron has a unit negative charge is subject to experimental deviations of later investigations (77) and draws support from the mechanistic assumptions (73) and hence may not be a well established fact.

The fact that the singlet e.s.r. signal is observed only in alkaline glasses at anout 6 - 10M [OH⁻] means that OH⁻ has an important role to play. OH⁻ has been proposed as structure former and can participate in hydrogen - bond formation and hence be accomodated in clusters of water (93). Due to the large difference in electronegativity of H and O of the water molecule the charge shall flow towards the more electronegative oxygen which should make the hydrogen bonding of the water molecules possible through the O and hence hydrogens shall be charge deficient in the cluster. Also OH⁻ can participate in hydrogen bonding due to this reason. Considering the single water molecule in the poly - water, OH⁻ shall block the entry of the excess charge on one H of the water molecule whereas the other shall be charge deficient.

When the electron is produced by photo - ionisation of the solute in water or by ionising water by high energy radiation, it can be accomodated on the charge deficient hydrogen of the water cluster resulting in the negative form of this cluster. The ionising radiati on (high energy gamma rays) or the U.V. light may make the medium inhomogeneous i.e. homogeneously distributed OH⁻ in the aqueous glassy matrix before its irradiation may result in the nonhomogeneous distribution of the OH⁻ after irradiation and the high concentration of OH⁻ in a particular region of the matrix shall block the entry of the charge in the nearest neighbouring hydrogen. The localisation of the charge on one hydrogen of the water molecules shall make the non - homogeneous part of the cluster as paramagnetic and the region where OH⁻ is not structure former may be diamagnetic because of the localisation of the charge in both hydrogens of the water molecule. When the electron is produced on irradiation, it will also help in making the distribution of the OH^- non - homogeneous due to subsequent repelling of the charge and polarisation of the medium. What happens when the matrix is warmed from the liquid N_2 temperature? The negatively charged water cluster recombines with the positively charged water cluster in the case of irradiations with Co^{60} gamma - rays and in the case of photo - ionisation of the solute with the u.v. light, it may recombine with the radical atom or other negatively charged speices e.g. in the case of I^- photolysis, the negatively charged cluster may recombine with I^o or its products with I^- if the paramagnetic cluster is observed on I^- photolysis and in the case of $Fe(CN)_6^{4-}$ photolysis, the negative cluster may destroy itself on its reaction with $Fe(CN)_6^{3-}$ or $Fe(CN)_6^{4-}$ dependent on their rate of reactions.

The distance of closest approach of the charge deficient hydrogen to the OH^- in the water cluster should be a function of temperature and $[OH^-]$ in the aqueous system. Thus at the same $[OH^-]$ as in the matrix but at room temperature the existence of a paramagnetic water cluster may not be possible and the near neutral water cluster can be easily assumed as diamagnetic due to the presence of extra charge produced by photolysis or radiolysis on the charge deficient hydrogens. This diamagnetic negatively charged cluster of water is proposed as the structure maker. Thus the band having $\lambda_{max} \simeq 7200 A^o$ obtained by Hart and Boag (15) is not due to hydrated electrons but due to this diamagnetic cluster of water. The band extends from shorter wavelength

down in the u.v. to the infra red region. Clearly, the transition
responsible for this band is np ← ns and the assymetry of the band and
its broadness at the absorption maxima can be ascribed to more than
one single thermal motion of the charge from the negative water cluster.
Recently Hart et al (105) have seen this band below 220 mμ also. Thus
the large scan of wavelength from far u.v. to the infra red due to this
negatively charged species may indicate that the species is polymeric
in nature. It is not within the framework of this thesis to propose
a definite structure of this polymer.

<u>Formation of structures with SF_6 and N_2O:</u>

The structure of the negative polymer of water with N_2O (structure
A) and with SF_6 (structure B) can be easily rationalised in terms of
the electrostatic model. Even though there may be resonance among
structures A, B and C of N_2O, but it is fair to assume that the negative end of the dipole of N_2O shall be present at oxygen because O
is electronegative as compared to N. When N_2O is in solution in water,
some perturbation by the solvent molecules on N_2O is a fair assumption
though there is no direct experimental evidence in this case as it is
in the case of SF_6(66). Assuming the possibility of hydrogen bonding
of the water cluster from the negative end of the dipole of the N_2O
molecule, the structure A is proposed as simply a weaker hydrogen bonding from the positive end of the dipole of the negatively charged
water cluster with the negative end of the dipole of the N_2O molecule.
Energetics may require that the extra charge present on the charge
deficient hydrogens of the water cluster shall still keep +ve end of

the dipole of this cluster at hydrogens only. The reverse case can also be considered i.e. the negatively charged water cluster shall have stronger dipole - dipole interactions (-ve end at O of the water cluster and +ve end of the N of the N_2O molecule) because the presence of extra charge on the charge deficient hydrogens of the water cluster has made the +ve end of the dipole weaker than when the charge was not present. Even if the extreme case of no solvent perturbation on N_2O is considered, the structure A may still be understood as the dipole - dipole interactions of the negative poly - water and the N_2O molecules.

The structure of the negative poly - water with SF_6 (structure B) can be understood in the same manner as with N_2O except that the positive end of the dipole shall lie on S atom and the -ve end on the F atom.

The electrostatic model of the structures with N_2O and SF_6 is speculative but perhaps this is the most reasonable approach that can be taken at the present time.

Validity of the existence of structures on the basis of the photo - chemical work:

It is fair to remark that $2537A^\circ$ light produces electron from I^- by the process of photo - ionisation because the ionisation potential of I^- is 3.08 ev (106) and $2537A^\circ$ light constitutes 4.8 e.v. energy. Thus, the assumption, that I^- is excited by $2537A^\circ$ light first and then the electron may be produced by the collision of water molecules with I^{-*} resulting in hydrated electrons, may not be valid. No formation of products from the photolysis of $5 \times 10^{-3}M$ KI solution indicates that the rate of breaking of the structure of the negative water cluster by

I^o may be faster than the rate of reaction of I^o with I^- but because the negative water cluster is seen by flash photolysis of I_{aq}^-, this argument may not be valid. Thus the products from the reaction of I^o with I^- i.e. I_2^- or I_3^- may act as structure breaker. This effect shall be more pronounced because there is no species present in the solution which can make structure with the negative water cluster. Perhaps, the most reasonable explanation is that on photo - ionisation of I^-, the survival of the electron depends on three competition processes:

(i)   Its rate of reaction with I^o

(ii)   the rate of reaction of I^o with I^-

(iii)   the rate of formation of the negative water cluster.

The increase of I_2 or N_2 yields with increasing $[N_2O]$ rules out mechanism (I) because if $\phi(I_2) = \phi(N_2) = \phi(\bar{e}_{aq})$ then the kinetics of the reactions postulated in the mechanism demand that at high $[N_2O]$ reaction (13) will be pseudo first order and $\phi(\bar{e}_{aq})$ should not be a function of $[N_2O]$. Any back reactions (e.g. reaction 19, 20) should also be prevented due to the low steady state $[\bar{e}_{aq}]$ at high $[N_2O]$. The yield of measured I_2 or N_2 is about 4.5×10^{-5}M for a quantum yield of ~ 0.23 in the present studies. Thus 10^{-3}M $[N_2O]$ should be more than enough to scavenge 10^{-4}M $\bar{e}_{aq}$. Clearly, the results shown in fig. 5 do not fit this mechanism. Another question that can be raised easily is that, is there any other mechanism that can fit the results? Believing the concept of the hydrated electron, mechanism (I) should be the most reasonable one that can be assigned to the results. But, no formation of N_2 using technique 2.9.1 a and the dependence of its yield on the technique used definitely rules out that reaction (13) is fast and diffusion controlled and favours the formation of N_2 in some post - irradiation manner dependent upon the technique used. Another strong experimental evidence that can be cited against the reaction (13) being fast is that when actinometry for 1849 A^o light is done with aqueous solutions of N_2O, N_2 is formed according to:

$$N_2O \xrightarrow[1849A^o]{h\upsilon} N_2 + O^{1D}$$

The yield of N_2 for this reaction is not dependent upon the techniques 2.9.1 a, b and c but is the same in all three cases. Thus, it is wise to understand that N_2 is not present in the solution in equilibrium with gas phase due to reaction (13) but is formed only when N_2O bubbles out of the solution at the time of analysis of the gaseous products. It is impossible that any paramagnetic species is stable for a long time at room temp. and thus the species that has formed structure (A) should be diamagnetic.

The formation of electron by the photo ionisation of I^- should polarise the medium due to charge - charge repulsion. The increase of $\Phi(I_2)$ or $\Phi(N_2)$ at increasing $[N_2O]$ perhaps indicates the non-homogeneous distribution of the radicals before their reactions. $[N_2O]$ is enough to prevent the reaction of I^o with the negative water cluster (i.e. I^o acting as structure breaker), the formation of structure (A) shall be favoured or structure (A) is formed in competition with

(i)   the rate of reaction of $\bar{e}$ with I^{o}

(ii)   the rate of reaction of I^{o} with I^{-}

(iii) the rate of formation of the negative water polymer and structure (A)

(iv) the rate of breaking of the structure of negative water cluster by I^{o} which shall be dependent on processes (i) and (ii) and (iii).

The yield of electrons that form negative water cluster can be calculated from the yield of the iodine assuming its formation due to reactions 12, 15 and 16 which is most reasonable. Only one electron can be produced from one I^{-} ion and thus reactions 12, 15 and 16 yield the stoichiometry:

$$\Phi(I_2) = \frac{1}{2}\Phi(\bar{e})$$

and because $\Phi(I_2) = \Phi(N_2)$ using technique 2.9.1 c (fig. 5), the stoichiometry

$$\Phi(N_2) = \Phi(I_2) = \frac{1}{2}\Phi(\bar{e})$$

represents the yield of electrons that form the negative water cluster and thus structure (A) at a particular $[N_2O]$. The dependence of the yield of N_2 on the technique employed means that the formation of N_2 is dependent on the rate and mechanism of the breaking of structure (A). No formation of N_2 in the first cycle using technique 2.9.1 a is due to the fact that at -196 °C, N_2O does not bubble out of the solution (freezing point of N_2O is -120°C). Perhaps during thawing in the second and third cycles some N_2O bubbles out of the solution due to disturbance of the equilibrium of the system and also some changes in the structure (A) may have been brought about due to freezing with liquid N_2 and thawing with hot water so as to form irreproducible yields of N_2.

The mechanism of the formation and breaking of structure (A) is subject to a further detailed investigation. However, it is proposed that the structure may break through the intermediate $N_2O^-_{aq}$.

SF_6 may form structure with the negative water cluster in a similar manner as with N_2O but the rates of making and breaking of the structure may be different in the two cases. The stoichiometry $\phi\frac{(F^-)}{6}$ may not represent the true quantum yield of the electrons that form the negative water polymer and the structure (B). If $\Phi(I_2) = \Phi(N_2) = \frac{1}{2}\Phi(\bar{e})$ represent the true quantum yield of the electrons that form structure (A) then $\Phi\frac{(F^-)}{6} = \frac{1}{2}\Phi(\bar{e})$ may be the quantum yield of electrons that form structure (B) between about 5×10^{-4} and 1.2×10^{-3}M SF_6 (figs. 3 and 7). However the quantum yield of the electrons that form structure (B) appear to be lower in comparison to structure (A) below about 5×10^{-4}M SF_6 accepting these stoichiometries (fig. 5). This may be attributed to the breaking of the structure (B) by a different mechanism at lower than the intermediate $[SF_6]$. The dropping of the quantum yield of fluoride after about 1.2×10^{-3}M SF_6 can also be interpreted perhaps because the stoichiometry $\Phi\frac{(F^-)}{6} = \frac{1}{2}\Phi(\bar{e})$ is not valid and the rate and mechanism of the breaking of the structure (B) may be different from the low and intermediate $[SF_6]$. However, the dropping of the quantum yield of fluoride can also be explained due to breaking of the structure of the negative water cluster by I^O which may be favoured by SF_6 by collision with I^O. As explained before, the radicals have a non-homogeneous distribution and in a particular region of the medium, the conditions may be favourable for the collision of I^O with SF_6 at high concentrations. The decreasing yield of fluoride with the increase in $[SF_6]$ and dose (figs. 6 and 8) supports the view that SF_6 helps I^O perhaps through collision in breaking the structure of the negative water cluster which is more pronounced as the $[I^O]$ increases with dose. The pronounced breaking of the negative water cluster by I^O or its products with I^- may also explain the negligible yield of iodine in all studies with $KI - SF_6$ systems.

The breaking of the structure (B) is proposed through the intermediate SF_6^-.

In the presence of small amounts of O_2, the abnormal increase in the fluoride yield may be due to O_2 helping SF_6 in breaking the structure of the negative water cluster and thus there may be no formation of structure (B) in the presence of O_2. O_2 can also be proposed as the structure breaker of the negative water cluster as O_2^- has been seen by pulse radiolysis (38) and O_2^- may act as structure breaker for structure (B). The same yield of N_2 and I_2 in the presence of O_2 in the photo - chemical studies of $KI - N_2O$ system (46) perhaps means that O_2 or O_2^- does not act as structure breaker for structure (A). No further point can be made regarding the effect of the O_2 and should be a subject of further detailed and interesting investigation.

From the discussion so far, it is obvious that the interpretation of the results of competition studies (figs. 9 - 19, tables 1 - 8) based on the kinetic mechanisms and derivations of the equations (9 - 11) from the reactions of SF_6 and N_2O with $\bar{e}_{aq}$ is not valid. It appears that structures (A) and (B) are formed in competition with each other depending on the concentrations of N_2O and SF_6 and the rate of formation of the structures is different from the rate of breaking of the structures. The difference in the degassed and non - degassed fluoride yields may be due to the different rate and mechanism operating for the formation of fluoride from the breaking of the structure (B) dependent upon the technique employed. The breaking of the structure (A) by technique 2.9.1 b may result in the intermediate product formation which can act as structure breaker for the structure (B) if both these structures are linked in a continuous net work. This is supported from the different degassed and non - degassed iodine yields (tables 2 - 7) in which the degassed iodine yields are lower than the non - degassed comparing at approximately the same $[\frac{SF_6}{N_2O}]$. The products resulting from the breaking of structure (B) by the structure breaking products from structure (A) may have a post - irradiation reaction with iodine. Perhaps SOF_2 or SO_3^{-2} is oxidised by I_3^- to yield SO_4^{-2} which means that SOF_2 and its precursors result from the post - irradiation breaking of structure (B). The difference in the non - degassed and degassed iodine yields may also be due to the different rate and mechanism of the breaking of the structure (B) and also of (A) by the two techniques.

At about 3.2×10^{-3} M N_2O, the unusually high nitrogen and iodine yield (fig. 18 and table 4) means that more electrons survive to form structure (A). This may be due to the effect of increased dose creating favourable conditions for the formation of structure (A) due to the pol - arisation of the medium.

The photochemical studies of KBr, KCl and KCNS each with SF_6 (figs. 20 - 26) support the view earlier expressed in the discussion for KI system that depending upon its concentration SF_6 helps the radicals e.g. Br^o, Cl^o, CNS^o perhaps through collision in breaking the negative water cluster. This can be clearly visualised from the fact that at 2.4×10^{-4} M SF_6, the fluoride yields are linear (figs. 20, 22, 25). However, it is also clear that the linearity or non - linearity of the fluoride yields as a function of the dose also depend upon the nature and size of the radical. In the case of KI - SF_6 photolysis (fig. 6), I^o acts as structure breaker of the negative water cluster in comparison to Cl^o and Br^o due to its biggest size.

The linearity of the fluoride yield as a function of dose should also be dependent upon the rates of the radical - ion reactions to form I_2^-, Cl_2^-, Br_2^-, and $(CNS)_2^-$. The absorption spectra of these ions is known from studies with fast reaction techniques (26, 76, 75). It appears that the reaction of CNS^o with CNS^- to form $(CNS)_2^-$ is much faster than the reactions of I^o, Cl^o and Br^o with I^-, Cl^- and Br^- respectively so that SF_6 at its high concentrations does not help CNS^o in breaking the structure of the negative poly - water.

be greatest and nuclear attraction of the charge to be photo - ionised is least in the case of I^- and thus it is possible to eject the electron from I^- with less energy. This should also be favoured due to the size of the I^- in comparison to Cl^- and Br^-.

A model for the quantum yield of electrons to form the negative water cluster in the case of photolysis of halides $(Cl^- > Br^- > I^-)$ is proposed as follows:

(i) Due to the biggest size of I^o, it will act as the best structure breaker of the negative water polymer and hence the measured quantum yield of electrons forming the negative water cluster from the photolysis of I^- shall be least.

(ii) The reaction of I^o with I^- to form I_2^- may be slow as compared to the reactions of Br^o and Cl^o with Br^- and Cl^- to form Br_2^- and Cl_2^- respectively. The least measured quantum yield of the electrons to form the negative water cluster in the case of photolysis of I^- may also be favoured due to structure breaking properties of I_2^- and I_3^- . The homogeneous kinetics to determine the rates of the radical - ion reactions may not be valid due to the non - homogeneous distribution of the rad-icals.

This model is speculative and is subject to further investigations.

The large quantum yields of F^- on the photolysis of CN^- is a radical departure from the other systems and may indicate a chain reaction, (table 9). The post - irradiation decrease in the fluoride yields prov-ides an evidence for a back reaction involving F^-. The structure (B) is

Drawing analogy from the KI - SF_6 system in which $\Phi\frac{(F^-)}{6} = \frac{1}{2}\Phi(e)$ is proposed as the quantum yield of electrons that form the negative water cluster, the same may be true in the case of KCl - SF_6, KBr - SF_6 and KCNS - SF_6 systems. Thus the stoichiometry $\Phi\frac{(F^-)}{6} = \Phi(\bar{e}_{aq})$ based on the reaction mechanisms of the hydrated electron can simply be changed to $\Phi\frac{(F^-)}{6} = \frac{1}{2}\Phi(\bar{e})$ for comparison of the quantum yields in the figures. It has generally been agreed upon so far that the quantum yield of hydrated electrons from the halides varies in the order $Cl^- > Br^- > I^-$. Discarding the erroneous concept of hydrated electron, it may be agreed upon that it is possible to propose the same thing that the quantum yield of electrons that form the negative water cluster vary in the order $Cl^- > Br^- > I^-$.

The electron shall be produced from Cl^- and Br^- by the process of photo - ionisation by $1849A^\circ$ light. The I.P. (ionisation potential) of Cl^- and Br^- is 3.62 and 3.38 e.v. (106) and $1849A^\circ$ light constitutes 6.6 e.v. energy. It may be argued that why $2537A^\circ$ light which constitutes 4.8 e.v. energy should not photo - ionise Cl^- and Br^-? The photo - ionisation of the halide should depend upon the interelectronic repulsion, size of the ion and nuclear attraction of the charge to be photo-ionised. Ionisation potential may take into consideration these three factors but how valid are the theoretical calculations and the estimation of the ionisation potentials from the gas phase spectra of the anion (106) which can correlate with the ionisation potential of the anion in the solution is subject to controversy due to solvent perturbations on the anion. Obviously the interelectronic repulsion should formed from the electrons produced by the photolysis of CN^-_{aq} but the structure may be unstable due to the abstraction of fluorine atom from SF_6 by CN^o as proposed in the following reaction scheme:

$$CN^o + SF_6 \longrightarrow FCN + SF_5 \quad\cdots\cdots\cdots\cdots (39)$$

$$SF_5 + CN^- \longrightarrow CN^o + SF_4 + F^- \quad\cdots\cdots\cdots (40)$$

$$SF_4 + H_2O \longrightarrow SOF_2 + 2F^- + 2H^+ \quad\cdots\cdots\cdots (41)$$

$$SOF_2 + 2H_2O \longrightarrow SO_3^{-2} + 2F^- + 4H^+ \quad\cdots\cdots\cdots (42)$$

$$CN + CN \longrightarrow (CN)_2 \quad\cdots\cdots\cdots\cdots\cdots (43)$$

$$(CN)_2 + F^- \longrightarrow FCN + CN^- \quad\cdots\cdots\cdots\cdots (44)$$

Reactions (39) and (40) provide the chain with reactions (43) being the terminating step and reaction (44) accounting for the post - irradiation decrease in the F^- yield.

This mechanism is novel in that it proposes the fluorine atom abstraction reaction (39). It is known that H atoms do not react with SF_6 in spite of the fact the H - F bond (145K. cal. /mole) is stronger than the S - F bond (76K.cal./mole) (87) and the reaction should be exothermic. In comparison, the fluorine - cyanobond is just 110K.cal./mole (88). Thus, the reaction must be kinetically rather than thermo-dynamically controlled.

The large quantum yield of fluoride in the photolysis of CN^- - SF_6 system can also be explained due to CN^o helping SF_6 in breaking the negative water cluster without fluorine atom abstraction. Every collision of CN^o with SF_6 may yield the intermediates of the breaking of the structure (B) through a complex mechanism. Pulse radiolysis and Flash photolysis should make an interesting study of the CN^- - SF_6 system to determine the intermediates resulting from the breaking of the structure (B).

A further application and test of the utility of SF_6 was made in the photolysis of $Fe(CN)_6^{4-}$. Whereas this system has been studied by several others (73, 77 - 79), there is a marked disagreement in the so called $\phi(e_{aq}^-)$. This also gives a strong support to the chemistry involved in the techniques of analysis of the products. S. Ohno (73, 79) has used technique 2.9.1 c for the analysis of N_2 and measured $Fe(CN)_6^{3-}$ in the spectrophotometric cell without opening it just after photolysis of the $Fe(CN)_6^{4-}$ - N_2O system. The kinetic analysis of the competition plots resulting from the reactions of N_2O and other solutes with e_{aq}^- has been done only with the yield of $Fe(CN)_6^{3-}$. No attempt has been made to do the kinetic analysis with the N_2 yields and the mechanisms proposed (73, 79) demand that the rate constant ratio should be the same considering the kinetic analysis of both N_2 and $Fe(CN)_6^{3-}$ yields. However, it has been mentioned in the experimental section (73) that N_2 was measured. The value of $\phi(e_{aq}^-) = 0.35$ was obtained from the yield of $Fe(CN)_6^{3-}$.

Dainton and Airey (77) and Shirom and Stein (78) used similar techniques of analysis (2.9.1 b) of N_2 resulting from the proposed and well accepted reaction of $\bar{e}_{aq}$ with N_2O but obtained different experimental numbers. $\Phi(\bar{e}_{aq}) = 0.66$ (77) and $\Phi(\bar{e}_{aq}) = 1.0$ (78) were obtained from the N_2 yields. Dainton and Airey (77) obtain $\Phi(\bar{e}_{aq})$ $= 0.66$ from the $Fe(CN)_6^{3-}$ yield also proposing a mechanism similar to S. Ohno (73, 79). The different so called $\Phi(\bar{e}_{aq})$ in each case leads credibility to the fact that the erroneous concept of the hydrated electron and its kinetic reactions with the scavengers is not valid. It may be implied that Dainton and Airey measured the yield of $Fe(CN)_6^{3-}$ after the analysis of N_2 (technique 2.9.1 b) and the increased yield over that measured by S. Ohno may be due to the reaction of the $Fe(CN)_6^{4-}$ with the intermediate product resulting from the breaking of structure (A).

The increase in the quantum yield of fluoride and $Fe(CN)_6^{3-}$ with $[SF_6]$ is due to the formation of negative water cluster and structure (B) in competition with the processes of structure breaking of the negative water cluster by the structure breaking products resulting from the photolysis of $Fe(CN)_6^{4-}$ (figs. 27, 28). Due to the high concentrations and fast reactions, the rates of the structure breaking reactions of $Fe(CN)_6^{4-}$ and $Fe(CN)_6^{3-}$ may give the plateau value of the $Fe(CN)_6^{3-}$ and fluoride yields if the rate of the formation of structure (B) is slower than the rates of the structure breakers (figs. 27 - 30). The continuous decrease of the $Fe(CN)_6^{3-}$ yield with dose (fig. 30) supports the view that $Fe(CN)_6^{3-}$ acts as structure breaker of the negative water cluster to yield $Fe(CN)_6^{4-}$ or may also give the formation of monoaqu o complex

$[Fe(CN)_5OH_2^{3-}]$ and the CN^- ion which suggests that the apparent yield

of $Fe(CN)_6^{3-}$ should be corrected for the yield of the monoaquo complex

in all cases. The small of HCN gas was observed (fig. 29) at higher

irradiation times (5 - 10 minutes) which may be due to the hydrolysis

of CN^- and naturally monoaqu o complex may also have been formed. The

absorption spectra of the monoaqu o complex overlaps that of the $Fe(CN)_6^{3-}$

hence an attempt could not be made to correct the apparent yield of

$Fe(CN)_6^{3-}$.

In the competition studies with $Fe(CN)_6^{4-}$ system (fig. 16),

the $\Phi[Fe(CN)_6^{3-}]$ remains constat. This can be interpreted as the apparent

yield of $Fe(CN)_6^{3-}$ resulting from the competition processes of the for-

mation of structures (A) and (B) and the structure breaking of the neg-

ative water cluster by $Fe(CN)_6^{3-}$ and $Fe(CN)_6^{4-}$.

The quantum yield of electrons that form the negative water cluster

and structure (B) may not be determined from the stoichiometry

$$\Phi \frac{(F^-)}{6} = \frac{1}{2} \Phi(\bar{e})$$

This stoichiometry may be relevant only when the N_2 yield is measured

by the photolysis of $Fe(CN)_6^{6-}$ - N_2O system by $2537A^\circ$ light by technique

2.9.1 c. Drawing analogy from the iodide system and assuming the similar

mechanisms of the breaking of structure (B) in the presence of iodide

or $Fe(CN)_6^{4-}$ or the products of the photolysis of these systems, $\Phi \frac{(F^-)}{6} =$

$\frac{1}{2} \Phi (\bar{e})$ may give the quantum yield of the electrons that form structure (B).

Chapter - VI

Discussion

Radiation - chemical results

VI - Discussion

The primary products presently accepted in the radiolysis of water are given on the right hand side of the following equation:

$$H_2O \xrightarrow{\hspace{2cm}} H_2, H_2O_2, H, OH, \bar{e}_{aq}, H_3O^+_{aq} \quad \cdots \cdots \cdots (45)$$

The radiolysis of water has commonly been described in terms of the spur which contains about four H_2O molecules decomposed in clusters within a radius of $20A^o$ to give $\bar{e}_{aq}$, OH, and H_3O^+. The reactions which give the primary products are usually described as follows:

$$\bar{e}_{aq} + \bar{e}_{aq} \rightarrow H_2 \quad \cdots \cdots \cdots \cdots \cdots \cdots \cdots (46)$$

$$\bar{e}_{aq} + H_3O^+ \rightarrow H \quad \cdots \cdots \cdots \cdots \cdots \cdots \cdots (47)$$

$$\bar{e}_{aq} + H \rightarrow H_2 \quad \cdots \cdots \cdots \cdots \cdots \cdots \cdots (48)$$

$$\bar{e}_{aq} + OH \rightarrow OH^- \quad \cdots \cdots \cdots \cdots \cdots \cdots (49)$$

$$H + H \rightarrow H_2 \quad \cdots \cdots \cdots \cdots \cdots \cdots \cdots (50)$$

$$OH + OH \rightarrow H_2O_2 \quad \cdots \cdots \cdots \cdots \cdots \cdots (51)$$

$$H + OH \rightarrow H_2O \quad \cdots \cdots \cdots \cdots \cdots \cdots (52)$$

No molecular products were observed by the steady - state radio-lysis of degassed water both at low and high dose which means that the intermediates formed as a result of the action of ionising radiation on water recombine to yield water. Interpreting the spectra observed by Hart and Boag (15) by the pulse radiolysis of degassed water as due to the negative water cluster perhaps indicates the presence of the corr-esponding positive water cluster also and the two recombine in about microseconds in the absence of any structure forming species in water. H and OH radicals are proposed as structure breaker of the negative water cluster. The rate of structure breaking by these radicals should be extremely fast which may demand that the negative water cluster should not be seen within the time resolution of the optics employed in pulse radiolysis. The intermediates may be non - homogeneously distributed and H and OH radicals which act as structure breakers of negative water cluster must have been produced in a region different from the one where negative water cluster has been seen.

$G(H_2) = 0.45$ observed in the presence of millimolar solution of KI is in agreement with the literature value. I^- is a good scavenger of OH radicals and thus the back reactions of H or H_2 with the inter-mediates present in water may not take place which may increase the possibility of reaction (50). The increase of $G(H_2)$ to a value 0.66 ± 0.06 (tables 10, 13, 18, 19, 22 and fig. 33) in the presence of struc-ture forming N_2O and SF_6 favours the view that only H atom is the pre-cursor to molecular H_2. N_2O and SF_6 form structures (A) and (B) respec-tively and thus H atoms which were taking part in the structure breaking of the negative water cluster appear as molecular H_2 due perhaps to the faster rate of the making of structures (A) and (B).

The low yield of $G(I_2) = 0.08 \pm 0.005$ produced as a result of $5 \times 10^{-3} M$ KI may be due to the structure breaking of the negative water cluster by I^o, I_2^- or I_3^-. However $G(I_2)$ increases considerably in the presence of structure making N_2O (tables 13, 14, 18, 19, 20, 21, 22 and fig. 33) whereas in the presence of SF_6 alone, $G(I_2)$ is nearly zero and there is a post - irradiation increase in $G(I_2)$ (tables 10, 11, 12). The increase in $G(I_2)$ in the presence of N_2O must be due to the increase in $G(OH)$. OH will oxidise I^- according to:

$$I^- + OH \rightarrow I^o + OH^- \qquad \ldots \ldots \ldots \ldots \ldots \ldots (53)$$

and I_3^- may be produced due to reactions (15) and (16). This increase in $G(OH)$ may be due to the faster rate of formation of structure (A) than the rate of breaking of the negative water cluster by OH and thus the nonhomogeneously distributed OH radicals show up as iodine. SF_6 may help I^o perhaps by collision in breaking the structure of negative water cluster which is in agreement with the negligibly small yield of iodine in the radiolysis of KI - SF_6 system.

Buxton and Dainton (18) propose that the formation of O_2 at the natural pH of the KI - N_2O system is due to the reaction:

$$H_2O_2 + I_2 \rightarrow O_2 + 2H^+ + 2I^- \qquad \ldots \ldots \ldots \ldots (54)$$

and thus $G_{H_2O_2} = G(O_2)$. From the present work, approximately zero

yield of O_2 using technique 2.9.1 a and different yields using tech-

niques 2.9.1 b and c suggest the formation of O_2 is due to some post -

irradiation reaction and since there is no formation of O_2 in the

photo - chemical studies with aqueous iodide system due to the absence

of the product H_2O_2, it is proposed that the formation of O_2 in radiation

chemical studies is due to the complex sequence of reactions of the inter-

mediate products resulting from the breaking of structure (A) with H_2O_2

I_2 and I^- and due to these reactions the degassed and non - degassed

iodine yields are different (tables 13 and 14). At natural pH, the

following thermal reactions of H_2O_2 with I^- have been studied previously

(108, 109):

$$H_2O_2 + I^- \longrightarrow IO^- + H_2O \quad \ldots \ldots \ldots \ldots \ldots \ldots (55)$$

$$IO^- + H_2O_2 \longrightarrow I^- + H_2O + O_2 \quad \ldots \ldots \ldots \ldots (56)$$

$$I_2 + H_2O_2 \longrightarrow 2H^+ + 2I^- + O_2 \quad \ldots \ldots \ldots \ldots (57)$$

but reactions (55) is slow which would not produce any significant

steady - state concentration of IO^- for the fast reaction (56) to occur.

As the pH of the KI solution is lowered, the reaction (55) is

accelerated and the reaction

$$H_2O_2 + 2H^+ + 3I^- \longrightarrow 2H_2O + I_3^- \quad \ldots \ldots \ldots \ldots (58)$$

progressively yields I_3^-, $k_{58} = 10.5$ M^{-1} sec.$^{-1}$ (109).

Buxton and Dainton (18) have reported that freezing the irradiated solution at $-80^{\circ}C$ accelerates the reaction (58). The yield of
degassed I_3^- reported in tables (10 - 12) in the KI - SF_6 radiolysis
may be due to post - irradiation reactions (55 - 58). O_2 or its precursors resulting from these reactions should act as structure breakers
for structure (B) leading to complex sequence of reactions. This
explanation is in agreement with no observed formation of O_2 in the
radiolysis of only KI - SF_6 and in competition studies (tables 13,
18, 19) where $G(O_2)$ decreases with the increase of $[SF_6]$ and decrease
of $[N_2O]$. The mechanism of the breaking of structures (A) and (B) is
dependent upon the techniques employed and hence the yields of the products shall differ in each case due to the post - irradiation reactions
of the system. Thus the degassed and non - degassed iodine yields differ
at about the same $[N_2O]$ in the radiolysis of KI - N_2O system (tables 13,
14), at about the same $[\frac{N_2O}{SF_6}]$ in the competition studies (tables 18, 19,
20, 21).

$G(H_2O_2)$ shows up as $G(O_2)$ due to its reactions with the products
resulting from the breaking of structure (A) both in competition studies
and in the radiolysis of KI - N_2O alone. However, in the case of the
radiolysis of KI - SF_6, H_2O_2 shows up quantitatively as iodine due to
reactions (55 - 58) and perhaps the oxidising intermediates resulting
from these reactions act as structure breakers for structure (B) and
this process of structure breaking shall lower the pH of the system due

to the yield of fluoride from the breaking of structure (B) and due to the lowering of pH reaction (58) may be accelerated. It is not understandable as to why freeze thaw cycle should accelerate the reaction. It cannot be due to the bubbling out of SF_6 from the solution and thus facilitating in the breaking of structure (B) because the freezing point of SF_6 is $-59^{\circ}C$.

The negligible amount of $G(I_2)$ in table II indicates that the mechanism of breaking of the structure (B) in the degassed and non - degassed cases is different. However the summation of $G(I_2)$ and $G(H_2O_2)$ is the same in the degassed and non - degassed cases (tables 11, 12) which clearly indicates that the increased amount of iodine produced in the degassed case (tables 10, 12) is due to the reactions of H_2O_2 which shows up quantitatively as I_3^-. If H_2O_2 is involved in the conversion of SOF_2 resulting from the breaking of structure (B) to SO_4^{-2}, its decrease due to such reactions should be the same in both degassed and non - degassed cases.

The positive water cluster is proposed as the precursor to the formation of H_2O_2. In the absence of the structure forming N_2O and SF_6, the positive water cluster recombines with the negative water cluster to yield the neutral water cluster. If the rate of the formation of structures (A) and (B) is faster than the rate of the recombination of positive and negative water clusters, then the positive water cluster may recombine with the neutral water cluster to yield H_2O_2. Anbar et al (107) noted the apparent impossibility of reconciling reaction (51) with the effect of adding $H_2^{16,16}O_2$ to $H_2^{18}O$, with $G(H_2^{18,18}O_2)$ decreasing and $G(H_2^{16,18}O_2)$ increasing while $2G(H_2^{18,18}O_2)$ + $G(H_2^{16,18}O_2)$ is

constant (107). The results favour the formation of H_2O_2 from the

recombination of positive and neutral water cluster. The reactions (59 - 62)

$$H_2O^{18} \wedge\!\!\wedge\longrightarrow H_2O^{18}{}^+ + H_2O^{18}{}^- \quad \ldots\ldots\ldots\ldots\ldots (59)$$

$$H_2O^{18}{}^+ + H_2O^{18} \longrightarrow H_2^{18}O_2^{18} + 2H^+ \quad \ldots\ldots\ldots (60)$$

$$H_2O^{18}{}^+ + H_2^{16,16}O_2 \longrightarrow H_2^{16}O_2^{18} + H_2^{16}O^+ \quad \ldots\ldots (61)$$

$$H_2O^{16}{}^+ + H_2O^{18} \longrightarrow H_2^{16,18}O_2 + 2H^+ \quad \ldots\ldots (62)$$

(The circles with the positive and negative charges represent the water

in the form of cluster with the respective charges)

may explain the increase of $G(H_2^{16}O_2^{18})$ by the addition of increasing

concentrations of $H_2^{16,16}O_2$ to $H_2^{18}O$ and clearly $2G(H_2^{18,18}O_2)$ + $G(H_2^{16}O_2^{18})$ should

be a constant. Clearly, the increase in the $[H_2^{16,16}O_2]$ favours

reactions (61) and (62) over (60).

The results of the competitions studies (tables 15 - 17 figs.

34 - 41) can be explained in terms of the post - irradiation breaking

of structures (A) and (B) and not the usual kinetic interpretation.

It can be easily seen that the rate constant ratio $k_{e + SF_6}/k_{e + N_2O}$

is not the same from the fluoride and N_2 analyses and yield of so called

hydrated electrons is dependent on the $[N_2O]$ and $[SF_6]$ and their reactions

with them are pseudo first order. The summation of $G(N_2) + G\dfrac{(F^-)}{6}$ is not a constant (table 16, 17 figs. 34 - 41) and increases with the

increase in $[N_2O]$ and or $[SF_6]$. Also $G(N_2)$ increases with the increase

in $[N_2O]$ in the radiolysis of KI - N_2O system (table 13). It is gen-

erally agreed upon that $G(N_2)$ = 2.4 $\pm$ 0.1 = $G(\bar{e}_{aq})$ in the radiolysis

of water and the dependence of $G(N_2)$ on (N_2O) may be due to another

reducing species perhaps $H_2O.^*$ But the same type of dependence of

$\Phi(N_2)$ = $\Phi(I_2)$ on $[N_2O]$ in the photochemical studies (fig. 5) gives

a conclusive evidence that H_2O^* is either not formed in the radiolysis

of H_2O or its role is not significant at all. However, no formation of

N_2 in the first freeze - thaw cycle and irreproducible and much lower

N_2 yields in the second and third cycles using technique 2.9.1 a (table

22) as compared to the techniques 2.9.1 b and c may be interpreted

that N_2 is only formed as a result of the post - irradiation breaking

of structure (A) when N_2O bubbles out of the solution. All these for-

going arguments undoubtedly rules out any interpretation based on the

usual kinetic mechanisms of the so called hydrated electrons.

Further, that fluoride and N_2 are formed as a result of the

breaking of structures (B) and (A) can be seen from the pronounced

difference in the non - degassed and degassed fluoride yields in the

competition studies (table 16, figs. 36 - 38). Assuming the validity

of the kinetic reactions of $\bar{e}_{aq}$ with N_2O and SF_6, it is demanded that

when $N_2O \gg SF_6$, the F^- yield should be negligible which is not the

experimental fact from the present work. In table 17, the increase in

$[SF_6]$ increases the $G(N_2)$ which further violates any kinetic interpre-

tation of the reactions of the so called $\bar{e}_{aq}$. All these inconsistenties are significant and should be the basis of further investigations of the mechanism and the nature of structures (A) and (B).

The intercepts of the reciprocal plots (figs. 34 - 41) yield different values of $G(\bar{e}_{aq})$ which again violates the basis of the competition equations 9 - 11.

$G(N_2)$ = 4.25 $\pm$ 0.1, 3.0 $\pm$ 0.1 and 3.1 $\pm$ 0.1 obtained from the intercepts of figs. 39 - 41 has the same values as in the radiolysis of KI - N_2O system (table 13) at the same $[N_2O]$ as employed in the competition studies. This shows purely a concentration effect of SF_6 to reduce the amount of electrons forming structure (A), and structure (B) is formed in competition with structure (A).

The mechanisms of the breaking of structures (A) and (B) may be similar in the photo - chemical and radiation - chemical studies . Thus $G(N_2)$ = $\frac{1}{2}G(\bar{e})$ may be valid for the yield of electrons forming structure (A) using technique 2.9.1 c and the 30 % increased yield by using technique 2.9.1b may be due to breaking of structure (A) by a different mechanism. The changes brought about in the structure (A) by freezing at $-80^{\circ}C$ and the rate and manner of breaking different from the techniques 2.9.1 a and c are the subject of further detailed investigations. The same so called rate constant ratio $k_{e + SF_6} / k_{e + N_2O}$ using the $Fe(CN)_6^{4-}$ system in the radiation - chemical studies (60) and the photo - chemical studies from the present work (fig. 16) favours the view that similar mechanism for the breaking of structure (B) are involved. This explanation is also favoured from the photolysis and radiolysis of iodide system from the present work. However the stoichiometries, $\phi\frac{(F^-)}{6}$ =

$\frac{1}{2} \Phi (\bar{e})$ and similarly $G\frac{(F-)}{6} = \frac{1}{2}G(\bar{e})$ obtained from the non - degassed fluoride yield may give the yield of electrons forming structure (B) but are subject to further detailed investigations of the rate and mechanism of the breaking of structure (B).

Chapter - VII

General

VII - GENERAL

It has been the general practice in flash photolysis and pulse radiolysis to assume that the rate of the disappearance of hydrated electrons in presence of their scavengers is also the rate of their reactions and hence, using usual rules of kinetics, the rate constants of the reactions are calculated from the half life of the reactions. The reactions are usually fast $\sim 10^{10}$ M-1 sec^{-1} and diffusion controlled. In the steady - state photolysis and radiolysis the rate constants of these reactions are calculated from the kinetic analysis of the reaction products based on the most suitable mechanism of the hydrated electron reactions. However, very seldom there has been agreement between the rate constants calculated from flash photolysis or pulse radiolysis and the steady - state photolysis and radiolysis experiments. These inconsistenties are well known in the most simple systems i.e. competition measurements for $\bar{e}_{aq}$ by N_2O and H^+_{aq} (29) and by NO_3^- and H^+_{aq} (73, 110). Even when $[H^+_{aq}] \gg [N_2O_{aq}]$, N_2 is formed (29). It is generally understood that H_2 is formed by the bimolecular recombination of $\bar{e}_{aq}$ and in the presence of the hydrated electron scavenger, $G(H_2)$ is suppressed from that obtained in presence of OH radical scavenger only. However, there is no information in the literature whether H_2 is also formed at low or high dose by steady - state or pulse radiolysis of degassed water in the absence of the $\bar{e}_{aq}$ scavenger. If $\bar{e}_{aq}$ is the precursor to the formation of H_2, then H_2 must be formed by the radiolysis of the degassed water. Recent work by Mahlman and Sworski (110) creates serious doubts that $\bar{e}_{aq}$ is the precursor of molecular H_2. These authors also point

out that the absence of any pH effect for the yield of molecular H_2, G_{H_2}

(although many except a rather small pH dependence), indicates that $\bar{e}_{aq}$

is not a precursor of molecular H_2 and 'excited water,' H_2O^*, which has

been commonly assumed.

Achievements of the present work:

Perhaps the present research done by the author and his explan-

ations of it has created vast problems in radiation chemistry instead of sol-

ving any. From the stand point of an inquisitive curious mind which wan-

dered in the world of radiation - chemists for a short time, the proposal

of negative and positive water clusters and structures (A) and (B) may

seem most logical to some but may appear meaningless to the mind which

only believes in the molecular orbital (charge delocalisation) approach.

The solution of the problems of the structures with the negative water

cluster is not an easy one and will require a long, hard and carefully

thought of research work. The studies by the fast reaction techniques

of the O_2 containing systems (i. e. KI - SF$_6$ O_2) or the cyanide systems

(KCN - SF$_6$) should yield useful information as to the nature of inter-

mediates which may result by the breaking of structure (B). The author

has also seen the appearance of blue colour using technique 2.9.1b in

the radiolysis and photolysis of KI - SF$_6$ - N$_2$O system. The blue colour

is definitely not due to iodine at -80°C because by freezing a synthetic

solution of I_3^- under similar conditions, an orange brown colour was

noted. The observation of blue colour should be confirmed by others and

may be followed through the matrix - isolation technique. In some photo -
chemical studies, the spectra study of the structures (A) and (B)
at room temperature was tried in the u.v. and visible range but no information could be obtained from such studies.

Another important question is the nature of the structures (A) and
(B), whether they are formed as a result of any chemical bond which is
straight chain or cyclic or any other geometric configuration or they
are held together and or singly just through dipole - dipole interactions.

So far, it has not been possible to see the spectra of the intermediates N_2O^- and SF_6^- that have been proposed in the literature due to
the reactions of $\bar{e}_{aq}$ with N_2O and SF_6. From the present work and its
explanations, it should not be surprising that these intermediates are
not formed by the earlier proposed fast reactions of $\bar{e}_{aq}$ with N_2O and
SF_6 but may result by the post - irradiation breaking of the structures
(A) and (B). Thus, the study of the formation and breaking of structures
(A) and (B) should be extremely interesting and requires immediate attention of the courageous experts of radiation chemistry.

Chapter - VIII

Conclusion

VIII - Conclusion

Any suitable mechanism based on the fast and diffusion controlled reactions of $\bar{e}_{aq}$ with N_2O and SF_6 in the photo - chemical and radiation - chemical studies is unlikely. The misinterpreted band with $\lambda_{max} \backsimeq$ $7200A^o$ due to hydrated electrons may be due to the diamagnetic negative water cluster which forms structures with N_2O and SF_6. SF_6 is proposed as structure breaker relative to N_2O of the negative water cluster. The basis of the formation of the negative water cluster and the structures (A) and (B) is the charge localisation on the charge deficient hydrogens of the neutral water cluster based on electronegativity difference of the elements involved in the structures. The structures (A) and (B) break in the post - irradiation period and the rates and mechanisms of breaking may be dependent upon the technique employed for the formation of the products.

In radiation - chemical studies, H is proposed as the only precursor of molecular H_2 and the positive water cluster as the precursor of molecular H_2O_2. H and OH act as structure breakers of the negative water cluster. The rate of formation of structures (A) and (B) may be greater than the rate of breaking of the structure of the negative water cluster by H and OH and hence $G(H_2)$ and $G(I_2)$ increases on the addition of N_2O but because SF_6 may help I^o in breaking the structure of the negative water cluster, $G(I_2)$ is nearly zero in all KI - SF_6 studies.

O_2 or its precursor is proposed as the structure breaker of structure (B) and it is formed due to the reactions of the intermediate product resulting from the post - irradiation breaking of structure (A) with H_2O_2, I_2 and I^-.

The distribution of the radicals formed in the present photo - chemical and radiation - chemical studies may be non - homogeneous and thus the use of homogeneous kinetics for the radical - ion reactions may not be valid.

'Excited water H_2O^{*}', is either not formed in radiation - chemical studies or its role is not important in the chemical reactions.

The stoichiometries $\Phi(N_2) = \Phi(I_2) = \frac{1}{2}\Phi(e)$, $G(N_2) = \frac{1}{2}G(e)$, $\Phi\frac{(F^-)}{6} = \frac{1}{2}\Phi(\bar{e})$ and $G\frac{(F-)}{6} = \frac{1}{2}G(\bar{e})$ may give the measured yield of electrons forming the negative water cluster using technique 2.9.1 c for measuring N_2 from structure (A) and the non - degassed fluoride yield from structure (B).

SF_6 acts as a wall to the radicals e.g. I^o, Br^o, Cl^o, CNS^o and perhaps CN^o. The radicals act as structure breakers of the negative water cluster dependent upon their steady - state concentrations and $[SF_6]$.

Chapter - IX

References

IX - REFERENCES

(1) F. W. Spiers, Discussions Faraday Soc., $\underline{12}$, 13(1952).

(2) G. J. Hine and G. L. Brownell, 'Radiation Dosimetry,' Academic Press Inc., New York (1958).

(3) F. Hutchinson and E. Pollard, Vol. 1, Chapter I, 'Mechanisms in Radiology,' (M. Errera and A. Forssberg, Eds.) Academic Press Inc., New York (1961).

(4) Spinks, F. W. T., Woods, R.J., 'An Introduction to Radiation Chemistry,' p. 50, John Wiley and Sons, Inc., New York (1964).

(5) C. M. Davidson and R. D. Evans, Revs. Mod. Phys., $\underline{24}$, 79 (1952).

(6) A. O. Allen, 'The Radiation Chemistry of Water and Aqueous Solutions,' D. Van Nostrand Company, Inc., Princeton, New Jersey (1961).

(7) W. Duane and O. Scheuer, Le Radium, $\underline{10}$, 33(1913).

(8) H. Fricke and E. R. Brownscombe, Phys. Rev., $\underline{44}$ 240 (1933).

(9) P. Guenther and L. Holzapfel, Z. Phys. Chem., $\underline{44B}$, 374 (1939).

(10) H. Fricke, E. J. Hart and H. P. Smith, J. Chem. Phys., $\underline{6}$, 229(1938).

(11) H. A. Schwarz, J. P. Losee and A. O. Allen, J. Am. Chem.Soc., $\underline{76}$, 4693(1954).

(12) E. J. Hart and R. L. Platzman, Vol. 1, Chapter 2, 'Mechanisms in Radiobiology,' (M. Errera and A. Forssberg, Eds.) Academic Press Inc., New York (1961)

(13) (a) R. H. Schuler and A. O. Allen, J. Chem. Phys., $\underline{24}$, 56(1956)

     (b) C. J. Hochanadel and J. A. Ghormley, J. Chem. Phys., $\underline{21}$,

880 (1953).

(c) R. M. Lazo, H. A. Dewhurst and M. Burton, J. Chem. Phys.,
22, 1370 (1954).

(14) A. H. Samuel and J. L. Magee, J. Chem. Phys., 21, 1080 (1953).

(15) J. W. Boag and E. J. Hart, J. Am. Chem. Soc., 84, 4090 (1962).

(16) G. R. Freeman, J. Chem. Phys., 46, 2822 (1967).

(17) J. C. Russell and G. R. Freeman, 48, 90 (1968).

(18) Buxton, G. V. and Dainton, F. S., Proc. Roy. Soc. A., 287,
427 (1965).

(19) Dainton, F. S. and Watt, W. S., Proc. Roy. Soc. A., 275, 447
(1963).

(20) Dainton, F. S. and Rumfeldt, R., Proc. Roy. Soc. A., 287, 444
(1965).

(21) Dainton, F. S. and Rumfeldt, R., Proc. Roy. Soc. A., 298, 239
(1967).

(22) Dainton, F. S., Chapter on 'The Chemistry of the Electron' in
'Fast Reactions and Primary Processes in Chemical Kinetics,'
p. 185 (and the references therein), Interscience Publishers,
New York (1967).

(23) Dainton, F. S., J. Pure and Applied Chemistry p. 15 (1967).

(24) E. Hayon, Trans. Faraday Soc., 60, 723 (1964).

(25) E. Hayon, Trans. Faraday Soc., 60, 734 (1964).

(26) Langmuir, M. E., Hayon, E., J. Phys. Chem., 71, 3808 (1967).

(27) Haissinsky, M., J. Chim. Phys., 60, 402 (1963).

(28) Haissinsky, M. and Patigny, P., J. Chim. Phys., 59, 675
      (1962).

(29) Head, D. A. and Walker, D. C., Can. J. Chem., 45, 2051 (1967).

(30) W. M. Hickam and R. E. Fox, paper presented before the 3rd
      annual meeting of the ASTM Committee E-14 on Mass Spectrometry,
      San Francisco, May 1955; Phys. Rev., 98, 557 (1955).

(31) Ahearn, A. J. and N. B. Hannay, J. Chem. Phys., 21, 119 (1953).

(32) Hickam, W. M. and Fox, R. E., J. Chem. Phys., 25, 642 (1956).

(33) Fox, R. E. and Curran, R. K., J. Chem. Phys., 34, 1595 (1961).

(34) Thoman, J. K., Gordon, S., Hart, E. J., J. Phys. Chem., 68,
      1524 (1964).

(35) Fox, R. E. and Curran, R. K. J. Chem. Phys., 34, 1590 (1961).

(36) Warman, J. M., Fessenden, R. W., J. Chem. Phys., 49, 4718 (1968).

(37) Stockdale, J. A. D., Christophorou, L. G., J. Chem. Phys., 48,
      1956 (1967).

(38) Czapski, G. and Dorfman, L. M., J. Phys. Chem., 68, 1169 (1964)

(39) Collinson, E., Dainton, F. S., Smith, D. R., Taxuke, S., Proc.
      Chem. Soc., 140 (1962).

(40) Czapski, G., Schwarz, H., J. Phys. Chem., 66, 471 (1962)

(41) Platzman, R. L. and Franck, J., Z. Physik, 138, 411 (1954).

(42) Anbar, M., Hart, E. J., J. Phys. Chem., 69, 1244 (1965).

(43) Weiss, J. J. in 'The Chemistry of Ionisation and Excitation'
      ed. by G. R. A. Johnson and G. Scholes, Taylor and Francis,
      London, p. 17 (1967).

(44) Hughes, G., Roach, R. J., Chem. Comm., 600 (1965)

(45) Walker, D. C., Can. J. Chem., 44, 2226 (1966)

(46) Dainton, F. S., Logan, S. R., Proc. Roy. Soc., London,A, 287, 281 (1965)

(47) Jortner, J., Ottolenghi, M. and Stein, G., J. Phys. Chem., 66, 2029, 2037, 2042 (1962)

(48) Jortner, J., Ottolenghi, M., Stein, G., J. Phys. Chem., 68, 247 (1964)

(49) Dainton, F. S., Sills, S. A., Nature, 186, 879 (1960)

(50) Dainton, F. S. Fowles, P., Proc. Roy. Soc., London, A, 287, 312 (1965)

(51) Stein, G., Treinin, A., Trans. Faraday Soc. 55, 1087 (1959)

(52) (a) Smith, M., and Symons, M. C. R., Discussions Faraday Soc., 24, 206 (1957)

       (b) Stein, G., Treinin, A., Trans. Faraday Soc., 55, 1091 (1959)

(53) (a) Smith and Symons, Trans. Faraday Soc., 54, 338, 346 (1958)

       (b) Griffiths and Symons, Trans. Faraday Soc., 60, 1125 (1960)

(54) G. Stein and A. Treinin, Trans. Faraday Soc., 56, 1393 (1960)

(55) R. L. Platzman and J. Franck, 'Farkas Memorial Volume.' Jerusalem, p. 21, (1952)

(56) Blandamer, Griffiths, Shields and Symons, Trans. Faraday Soc., 60, 1524 (1964)

(57) Chadwell, H. M., Chem. Revs., 4, 375 (1927)

(58) Hughes and Willis, Disc. Faraday Soc., 36, 214 (1963)

(59) Dainton, F. S., Gibbs, A., and Smithies, D., Trans. Faraday Soc. 62, 3170 (1966)

(60) Asmus, K. D., Fendler, J. H., J. Phys. Chem., 72, 4285 (1968)

(61) G. R. A. Johnson and J. M. Warman, Trans. Faraday Soc., 61, 1709 (1965)

(62) J. M. Warman, K., D. Asmus and R. H. Schuler, Advance Chem. Series, 00, 0000 (1968) and references therein.

(63) G. R. A. Johnson and H. Simic, J. Phys. Chem., 71, 2775 (1967)

(64) Hatchard, C. G., and Parker, C. A., Proc. Roy. Soc. A., 235, 518 (1956)

(65) Dainton, F. S., Fowles, P., Proc. Roy. Soc. A., 287, 295 (1965)

(66) Friedman, H. L., J. Am. Chem. Soc., 76, 3294 (1954)

(67) Solubilities of Inorganic and Metal - organic Compounds, IV ed., Vol. II, ed., W. F. Linke, Amer. Chem. Soc. p. 794 (1965)

(68) Awtrey, A. D., Connick, R. E., J. Am. Chem. Soc., 73, 1842 (1951)

(69) Durst, R. A., Taylor, J. K., Anal. Chem., 39, 1483 (1967)

(70) K. Srinivasan and G. A. Rechnitz; Anal. Chem., 40, 509 (1968)

(71) Allen, A. O., Hochanadel, C. J., Ghormley, J. A., and Davis, T. W., J. Phys. Chem., 56, 575 (1952)

(72) Rumfeldt, R. C., Batiste, J., Unpublished Work

(73) Ohno, S., Bulletin Chem. Soc., Japan, 40, 1770, 1776 (1967)

(74) Anbar, M., Myerstein, D., and Neta, P., J. Phys. Chem., 68, 2967 (1964)

(75) Baxendale, J. H., Scott, D. A., Chem. Commun. 14, 699 (1967)

(76) Matheson, M. S., Mulac, W. A., Rabini, J. A., J. Phys. Chem., 67, 2613 (1963)

(77) Airey, P. L., Dainton, F. S., Proc. Roy. Soc., A, 291, 340, 478 (1966)

(78) M. Shirom and G. Stein, Nature, 204, 778 (1964)

(79) S. Ohno, Bull. Chem. Soc., Japan, 40, 1779 (1967)

(80) W. M. Latimer and W. H. Rodebush, J. Am. Chem. Soc. 42, 1419 (1920)

(81) Gordon, S., Hart, E. J., Matheson, M. S., Rabani, J. and Thomas, J. K., Discuss. Far. Soc., 36, 193 (1963)

(82) L. I. Grossweiner and M. S. Matheson; J. Phys. Chem. 61, 1089 (1957)

(83) J. K. Thomas, Trans. Far. Soc., 61, 702 (1965)

(84) Edgecombe, F. H. and Norrish, R. G. W., Proc. Roy. Soc. A, 253, 154 (1959)

(85) Dainton, F. S. and Logan, S. R., Trans. Far. Soc., 61, 715 (1965)

(86) Hamill, W. H., J. Phys. Chem., 75, 1341 (1969)

(87) H. L. Roberts; Quart. Reviews, p. 30 (1960)

(88) Douglas, D. Davis and H. Okabe; J. Chem. Phys., 49, 5526 (1968)

(89) Cook, E., Siegel, B., J. Inorg. Nucl. Chem., 29, 2739 (1967)

(90) W. H. Barnes, Proc. Roy. Soc. (London) A125, 670 (1929)

(91) L. Pauling, The nature of the Chemical Bond, 3rd edition, p. 468.

(92) H. S. Frank, Proc. Roy. Soc. (London) A 247, 481 (1958)

(93) H. S. Frank and W. Y. Wen, Discussions Faraday Soc., 24, 133 (1957).

(94) Nemethy, G, and Scheraga, H. A., J. Chem. Phys. 36, 3382 (1962)

(95) Reviewed by G. C. Pimentel and A. L. McClellan, The Hydrogen
Bond (W. H. Freeman and Company, San Francisco, 1960), p. 214

(96) E. J. W. Verwey, Rec. Trav. Chim. 60, 887 (1941)

(97) K. Grjotheim and J. Krogh. Moe, Acta Chem. Scand. 8, 1193 (1954).

(98) Craig, Maccoll, Orgel and Sutton, J. Chem. Soc., London, 1954,
P.p. 352, 354

(99) Craig and Magnusson, ibid., 1956, p. 4895

(100) L. Pauling, 'The Nature of the Chemical Bond'. 3rd edition, p. 185

(101) R. G. Shulman, B. P. Dailey and C. H. Townes, Phys. Rev. 78, 145
(1950)

(102) D. Schulte, Frohlinde and K. Eiben; Z. Naturforschung, 17a, 445
(1962)

(103) F. S. Dainton, G. A. Salmon, and J. Teply. Proc. Roy. Soc. A,
286, 27 (1965)

(104) F. S. Dainton and G. A. Salmon, Proc. Roy. Soc. A, 285, 319 (1965)

(105) Hart, E. J., Christensen, H, Nilsson, G., J. Phys. Chem. 73, 3171
(1969)

(106) Berry, Reiman and Spokes, J. Chem. Phys., 37, 2278 (1962)

(107) M. Anbar, I. Pecht, and G. Stein, J. Chem. Phys., 44, 3635 (1966)

(108) Liebhafsky, H.A. and Mohammad, A., J. Am. Chem. Soc. 55, 3977
(1933)

(109) Liebhafsky, H. A., J. Am. Chem. Soc. 54, 3499 (1932)

(110) H. A. Mahlman and T. J. Sworski, 'The Chemistry of Ionisation and Excitation,' G. R. A. Johnson and G. Scholes, Ed., Taylor and Francis Ltd., London, 1967.

(111) Megregian, S., Anal. Chem., 26, 1161 (1954)

Chapter – X

Applications

POTENTIAL APPLICATIONS OF STABLE NEGAATIVE CHARGE WHICH IS KNOWN AS HYDRATED ELECTRON

The references covered in this book are up to the year 1969, when the thesis was written. The author has published this book in order to draw the attention of scientists of the world to use the experimental data covered in this book for future research. The author envisions three main applications of the stability of Hydrated Electron. Research experiments can be designed on the basis of stable negative charge to investigate the possibility of SUPERCONDUCTIVITY at room temperature. It has been reported in several publications in literature that Superconductivity of several solid superconductors is achieved at low temperatures such as at liquid nitrogen temperatures or higher. It has yet to be proven in a single case about the existence of superconductivity at room temperature. The evidence of the stability of negative charge under certain conditions as expressed in this thesis can be utilized for further experimentation to find the Superconductivity phenomena at room temperature.

Second application envisioned is for the experimentation on high energy density batteries for usages in cars, and other gadgets. Because in this thesis, there is evidence of the stability of negative charge, electrodes and electrolytes can be designed in several systems to provide high energy density batteries in deaerated electrolyte systems such as Lithium/Water, Zinc/Silver Oxide, Lithium/Conducting Polymers, Lithium/Sulfur Dioxide, Lithium/Mixture of Non-Aqueous and Aqueous Electrolyte. These are some examples of ambient temperature battery systems, which can be experimented based on the stability of hydrated electron.

Third application envisioned is for producing Polymers by the method of ANIONIC POLYMERZATION based on the stability of the negative charge and structure-breaking aspects by the Monomers of stable hydrated electron.

Chapter – XI

Publications

Indian Journal of Chemistry
Vol. 10, July 1972, pp. 718-723

Formation of Structures with the Negative Water Cluster, $n(H_2O)^2$: Part I—Photochemical Studies

R. K. SINGAL*

Polymer Institute, University of Detroit, Detroit, Mich. 48221

Manuscript received 22 *February* 1971; *accepted for publication* 13 *January* 1972

Photolysis of deaerated millimolar aqueous solutions of KI and $Fe(CN)_6^{4-}$ with 2537 Å light and of KCl, KBr and KCNS using 1849° light were carried out in the presence of N_2O and/or SF_6. The solution and gaseous products do not correspond to any suitable mechanism based on fast and diffusion controlled reactions of e_{aq}^- with N_2O and/or SF_6. It is proposed that (i) species e_{aq} is not structureless but has definite structure; (ii) this species has structure forming characteristics; (iii) molecules which form structures with this species can be classed as structure makers and which break this structure can be classed as structure breakers. Radicals are proposed as structure breakers only; (iv) N_2O and SF_6 are both structure forming molecules but SF_6 is structure breaker relative to N_2O. O_2 is proposed as structure breaker only. These hypotheses are justified from this study and the information available from the literature. The species proposed is diamagnetic negative water cluster (NWC) the basis of which and of its structures with N_2O and SF_6 is the localization of extra charge produced on photolysis of anions on the charge deficient hydrogens of the neutral water cluster due to electronegativity difference of the elements involved in the structures. The structures are proposed to break through their intermediates N_2O^- and/or SF_6^- in the post-irradiation period. SF_6 acts as a wall to the radicals, e.g., $Br°$, $Cl°$, $I°$, $CNS°$. The radicals have a non-homogeneous distribution and act as structure breakers of the NWC, depending on their steady state concentration and of $[SF_6]$.

Materials and Methods

The solutions were prepared in water which had been distilled from acid dichromate in a stream of oxygen over a hot platinum filament, condensed and redistilled from alkaline permanganate, and maintained under constant reflux over alkaline permanganate. The salts KCNS, KBr, KCl, KI, $K_4[Fe(CN)_6]$ were of reagent grade and used without further purification. The gases N_2O and SF_6 (Matheson Co.) were purified by repeated trap-to-trap distillation, stored in the vacuum system, and redistilled several times just prior to use.

Apparatus and procedures — The standard photolysis vessel consisted of a cylindrical quartz cell (20 mm in depth) to which was attached a 100 ml degassing bulb and a 10 mm path length spectrophotometer cell. The solutions were pipetted into the vessel through a side arm to which a stopcock could be joined for connecting to the vacuum system. After thorough deaeration, the solutions were equilibrated with the desired partial pressure of SF_6 or N_2O. Since the pressure of the additive gases was measured with a mercury manometer, the solutions were protected from contamination by mercury vapour by placing a trap (CO_2-acetone with N_2O and ice-salt slurry with SF_6) between the manometer and the reaction vessel.

Since SF_6 is only slightly soluble in H_2O (ref. 9), relatively high partial pressures were required to obtain millimolar concentrations and a high pressure cell was used for these experiments. This cell consisted of a heavy walled pyrex cup and side arm which were enclosed in a brass jacket. A quartz window was sealed on to the top of the cup by an O-ring and held tightly in place by a

I N the past decade, there has been considerable interest in the novel species, 'hydrated electron' (e_{aq}^-). Its reactions with the scavengers are generally considered fast ($\sim 10^{10}$ mole^{-1} litre^{-1} sec^{-1}) and diffusion-controlled. The species is short-lived (half-life $\simeq 300$ μsec) and is claimed to be paramagnetic[1-3]. Homogeneous kinetics has been used to determine its rate of reaction and the rate of disappearance of e_{aq}^- in the presence of excess [scavenger] (pseudo-first order reaction) is taken to be the rate of formation of products in steady state radiolysis or photolysis, pulse radiolysis or flash photolysis of aqueous solutions. The rates of reactions when studied both by conventional steady state techniques or fast reaction techniques, do not agree, even in the most simple systems, e.g. competition reactions of e_{aq}^- with H^+ and N_2O, reactions of e_{aq}^- with NO_3^- and other solutes[4-7].

SF_6 is known[8] to react rapidly with e_{aq}^- with the proposed stoichiometry

$$e_{aq}^- + SF_6 \rightarrow SO_4^{2-} + 6F^- + 8H^+ \qquad \dots(1)$$

In the past, N_2O has been used to a great extent in radiation-chemical studies to determine quantitatively e_{aq}^- yield. Since otherwise SF_6 is generally assumed to be inert, reaction (1) represents a useful alternate to the use of N_2O as a specific scavenger of e_{aq}^-. Present study was undertaken in order to investigate the utility of SF_6 as a scavenger of radicals produced on photolysis or radiolysis of KI system and the photolysis only of Cl^-, Br^-, SCN^-, $Fe(CN)_6^{4-}$.

---

* Present address: Research Department, Kerr Manufacturing Company, 28200 Wick Road, Romulus, Michigan 48174, USA.

threaded metal cap which screwed over the window on to the brass jacket. A stopcock with a threaded teflon tap was attached to the side arm to close the system. This cell could withstand at least 15 atm internal pressure without leaking. The gas to be added to the cell was first measured in a calibrated bulb then distilled into the reaction vessel and the concentration of the gas in solution calculated from the solubility of SF_6 and the PV data with due corrections made for the volume of the solution.

The photochemical lamp was a low pressure mercury grid lamp (model PCQ 011S, Ultraviolet Products, Inc., San Gabriel, California). For irradiations with the 2537 Å line the cells were immersed in a constant temperature water-bath (25° ± 0·5°C) with the radiation passing through a quartz window. When the desired line was 1849 Å, the cell was first brought to 25°C in the bath, dried, and irradiated. Both the cells contained magnetic stirring bars and all solutions were vigorously stirred during irradiation.

The intensity of the 2537 Å line was determined using the ferrioxalate actinometer[10], first with a water filter then replacing it with a degassed aqueous KI filter and measuring the difference. The light intensity calculated for the low pressure cell was $1·08 \times 10^{20}$ quanta/litre/min and for the high pressure cell was 8×10^{19} quanta/litre/min. A deaerated solution of N_2O was used as the actinometer for the 1849 Å line. Following the work of Dainton and Fowles[11] and assuming $\Phi°(N_2) = 1$ obtained by the direct photolytic decomposition of N_2O, a dose rate of $3·1 \times 10^{18}$ quanta/litre/min was calculated for the low pressure reaction vessel and $3·55 \times 10^{18}$ quanta/litre/min for the high pressure cell. The dose rates were calculated after correcting for the light fraction absorbed by N_2O and that absorbed by H_2O. The observed $\Phi(F^-)$ was also corrected for the light fraction absorbed by the

anions photolysed and that absorbed by water. For calculations, the same extinction coefficients at 1849 Å for Cl^- and Br^- were used as obtained by Dainton and Fowles and for SCN^-, a value of $1·63 \times 10^4 \pm 100$ litre mole^{-1} cm^{-1} was calculated from its spectra.

Product analysis — Three techniques were used for collecting the gases in the calibrated capillary. After photolysis, the solution was frozen with liquid nitrogen (technique-a), dry ice-acetone (technique-b), and no refrigerant (technique-c), and the gaseous products taken from the reaction vessel in three freeze-pump-thaw cycles were passed through traps at −196°C to remove condensable gases and collected into the pre-calibrated capillary tube by means of several cycles of the toepler pump during each freeze-pump-thaw cycle. Quantitative analysis of the gaseous products was done using a Varian model 1521-1B gas chromatograph with a molecular sieve 5A column and a thermal conductivity detector. The instrument was calibrated for the analysis of O_2 and N_2 using pure samples of each gas. N_2 was the only gaseous product in all the experiments and the reliable reproducibility was no better than ± 3-6%.

Solution products measured after the removal of gaseous products from the post-photolysis solutions are termed as 'degassed' and the ones measured just after the photolysis, i.e. without measuring the gaseous products, are termed as 'non-degassed'. I_2, F^-, SO_4^{2-}, $Fe(CN)_6^{3-}$ were the solution products measured. Monoaquo complex $Fe(CN)_5.OH_3^{3-}$ that may have formed in the photolysis of $Fe(CN)_6^{4-}$ system could not be measured as its spectra overlaps that of $Fe(CN)_6^{3-}$ (ref. 12). The total yield of I_2 was calculated using values of $2·64 \times 10^4$ mole litre^{-1} cm^{-1} for the decadic molar extinction coefficient of the I_3^- ion[13] and 742 moles litre^{-1} for the equilibrium constant $[I_3^-]/[I^-][I_2]$ (ref. 14). F^- yields were

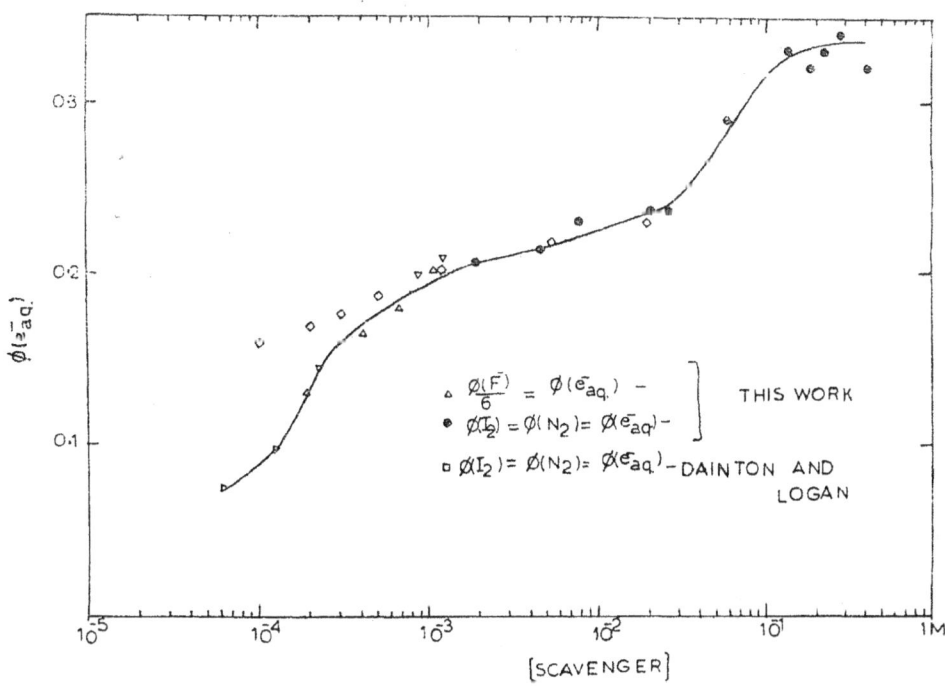

Fig. 1 — $\phi(e_{aq}^-)$ as a function of [scavenger] [Scavengers = SF_6 and N_2O]

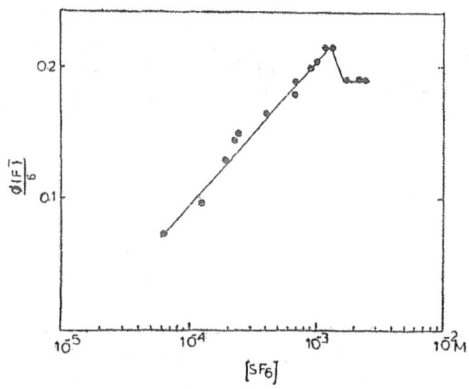

Fig. 2 — Quantum yield of F⁻ as a function of [SF₆] [Dose = 8×10^{19} quanta for SF₆ conc. $> 2 \cdot 4 \times 10^{-4}M$ and $1 \cdot 08 \times 10^{20}$ quanta for SF₆ conc. $\leqslant 2 \cdot 4 \times 10^{-4}M$]

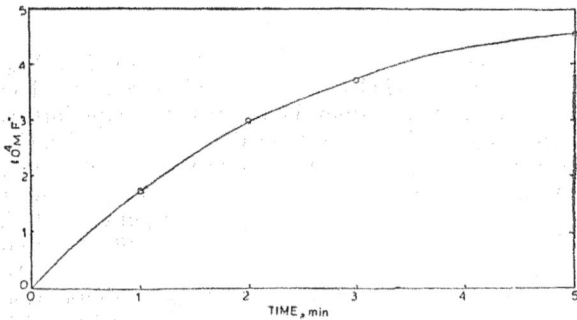

Fig. 3 — Yield of F⁻ as a function of irradiation time [SF₆ conc. $= 1 \cdot 2 \times 10^{-3}M$; and dose rate $= 8 \times 10^{19}$ quanta/min]

measured by an Orion Research Specific Ion Fluoride Electrode and the Orion Research Specific Ion Meter model 801. A calomel electrode was used as the reference in conjunction with the F⁻ electrode. Details of the procedure of calibration of F⁻ electrode are given in literature[8,15]. In the present study, the range of F⁻ yield measurement was kept between 5×10^{-5}-$10^{-6}M$ by proper dilutions of the photolyte so that pH of the unknown F⁻ solutions did not go below approximately 4·7. In the pH region of 4·5-7·5, the electrode is completely selective for F⁻ and measures the total [F⁻] of the unknown samples.

Quantitative analysis of the sulphate yield was not achieved due to its irreproducibility. For the analysis, some experiments were tried with nephlometric technique[16] but no meaningful result could be obtained. However, it was detected by a solution of BaCl₂.

Results

All experiments have been done at the natural pH of the deaerated solutions.

KI-N₂O system — A series of experiments were carried out using $5 \times 10^{-3}M$ KI solution containing different [N₂O]. The quantum yield of I₂ (non-degassed; measured in the side arm without opening the reaction vessel) and N₂ (technique-c) were equal and the results (Fig. 1) at $1 \cdot 5 \times 10^{-2}M$ [N₂O] are in agreement with Dainton and Logan[14]. The results (Fig. 1) at [N₂O] higher than $2 \cdot 6 \times 10^{-2}M$ have not been reported previously for photochemical studies.

However, the quantum yield of N₂ (technique-b) at $1 \cdot 5 \times 10^{-2}M$ N₂O was $0 \cdot 285 \pm 0 \cdot 005$ and that of I₂ (degassed) was $0 \cdot 265 \pm 0 \cdot 005$. These results are in agreement with Jortner *et al.*[17,18]. Some experiments were done using $5 \times 10^{-3}M$ solution of KI containing $1 \cdot 5 \times 10^{-2}M$ N₂O and measuring N₂ by technique-a, and I₂ after the N₂ analysis. The yields of N₂ were irreproducible and much lower than measured by techniques-b and c. In the first freeze-pump cycle, there was no N₂. The behaviour of the first cycle was always reproducible but not in the second and third cycles. The quantum yield of I₂ measured after the N₂ analysis was $0 \cdot 24 \pm 0 \cdot 01$.

KI-SF₆ system — At [SF₆] $= 2 \cdot 4 \times 10^{-4}M$, yield versus dose studies were done in the dose range $0 \cdot 54$-$5 \cdot 4 \times 10^{20}$ quanta showing the linearity in F⁻ yield to $1 \cdot 08 \times 10^{20}$ quanta and then continuously decreasing with increasing amount of dose. $\Phi(F⁻) = 0 \cdot 87 \pm 0 \cdot 03$ was calculated from the initial linear portion. Fig. 2 shows the variation of $\Phi(F⁻)$ with SF₆ and the maximum quantum yield of e_{aq}^-, assuming $\Phi(F⁻)/6 = \Phi(e_{aq}^-)$ achieved was only $0 \cdot 22$ at about $1 \cdot 2 \times 10^{-3}M$ SF₆ after which it appears to drop. Yield versus dose plot was attempted at $1 \cdot 2 \times 10^{-3}M$ SF₆ (Fig. 3) showing continuous curvature in F⁻ yield with increasing dose and from the initial slope of the curve, $\Phi(F⁻)/6 = 0 \cdot 225 \pm 0 \cdot 005$ was calculated. $\Phi(I_2)$ was found to be $0 \cdot 02 \pm 0 \cdot 02$ for all KI-SF₆ systems. In preliminary experiments with $5 \times 10^{-3}M$ KI solution containing $2 \cdot 4 \times 10^{-4}M$ SF₆, the quantum yield of F⁻ was much higher than reported in the results earlier. More thorough deaeration of KI solution (before the addition of SF₆) resulted in the reproducible yields of F⁻. The same trend of high F⁻ yield was noted in experiments with N₂O and SF₆.

KI-SF₆-N₂O systems — The chemical data obtained from such studies was analysed by well known reciprocal plots using the Eqs. (1)-(3) (see discussion for the sequence of reactions used for the derivation of these equations):

$$[\Phi(N_2)]^{-1} = [\Phi°(e_{aq}^-)]^{-1} + [\Phi°(e_{aq}^-)]^{-1} \cdot \frac{k_{e+SF_6}[SF_6]}{k_{e+N_2O}[N_2O]} \quad \cdots (1)$$

$$\left[\frac{\Phi(F⁻)}{6}\right]^{-1} = [\Phi°(e_{aq}^-)]^{-1} + [\Phi°(e_{aq}^-)]^{-1} \cdot \frac{k_{e+N_2O}[N_2O]}{k_{e+SF_6}[SF_6]} \quad \cdots (2)$$

$$[\Phi°(I_2)]^{-1} = [\Phi°(e_{aq}^-)]^{-1} + [\Phi°(e_{aq}^-)]^{-1} \cdot \frac{k_{e+SF_6}[SF_6]}{k_{e+N_2O}[N_2O]} \quad \cdots (3)$$

where $\Phi(N_2)$, $\Phi(F⁻)$ and $\Phi(I_2)$ are the measured yields of the products and $\Phi°(e_{aq}^-)$ is the total yield of hydrated electrons. The results for each system are summarized in Table 1. The experiments were carried out by fixing one [scavenger] and varying the other and the nitrogen, fluoride and iodine yields were analysed kinetically employing Eqs. (1), (2) and (3) respectively. From the best intercept of these plots, $\Phi(N_2)$ and $\Phi(F⁻)/6$ were determined and the slope of such plots gave the rate constant ratio k_{e+SF_6}/k_{e+N_2O} as shown in Table 1.

A perceptible difference in the non-degassed and degassed fluoride yields at about the same [scavenger] was observed in all the cases from approximately greater than $5 \times 10^{-3}M$ [N₂O] if [SF₆] is fixed and at about all [SF₆] if [N₂O] is fixed. The degassed fluoride yields are lower than the non-degassed yields and this behaviour is reversed in the radiation-

TABLE 1 — YIELDS AND THE RATE CONSTANT RATIOS IN THE KI-N$_2$O AND KI-SF$_6$ SYSTEMS

Dose (quanta)	$[SF_6] \times 10^4 M$	$N_2O \times 10^3 M$	Analysis of F$^-$ yield		Analysis of N$_2$ yield	
			$\dfrac{k_{e+SF_6}}{k_{e+N_2O}}$	$\dfrac{\phi(F^-)}{6}$	$\dfrac{k_{e+SF_6}}{k_{e+N_2O}}$	$\phi(N_2)$
1.08×10^{20}	1.58	1.73 — 8.6	4.7 ± 0.15[a]	0.225 ± 0.01[a]	—	—
1.08×10^{20}	1.58	1.73 — 8.6	4.08 ± 0.15[b]	0.225 ± 0.01[b]	19.7 ± 1	0.245 ± 0.01
2.16×10^{20}	1.58	1.73 — 8.6	3.9 ± 0.15[a]	0.238 ± 0.01[a]	—	—
2.16×10^{20}	1.58	1.73 — 8.6	3.5 ± 0.15[b]	0.238 ± 0.01[b]	21.5 ± 1	0.23 ± 0.01
1.08×10^{20}	0.51 — 1.9	3.46	4.8 ± 0.15[a]	0.238 ± 0.01[a]	—	—
1.08×10^{20}	0.51 — 2.22	1.73	5.3 ± 0.15[a]	0.238 ± 0.01[a]	—	—
1.08×10^{20}	0.51 — 1.9	3.46	4.2 ± 0.15[b]	0.238 ± 0.01[b]	20 ± 0.5	0.24 ± 0.005
1.08×10^{20}	0.51 — 2.22	1.73	4.0 ± 0.15[b]	0.238 ± 0.01[b]	18 ± 0.6	0.24 ± 0.005
8×10^{19}	0.41 — 17.7	2.4 — 22.7	3.9 ± 0.15[c]	0.238 ± 0.01[c]	—	—

(a) Analysis of non-degassed F$^-$ yields.
(b) Analysis of degassed F$^-$ yields.
(c) Analysis of F$^-$ yields obtained from high pressure studies.

chemical studies (unpublished data). Some competition experiments were also attempted at high pressures of SF$_6$ (more than 1 atm.) In the high pressure reaction vessel. The non-degassed F$^-$ yields were plotted which gave the rate constant ratio k_{e+SF_6}/k_{e+N_2O} (Table 1) lower than the value determined at low pressures of SF$_6$.

The N$_2$ yields (technique-b) were analysed kinetically. A marked difference is seen in the rate constant ratio determined by F$^-$ yields and that determined by N$_2$ yields (Table 1). Some competition experiments were also tried for technique-a for the measurement of N$_2$ and in all the cases the yield of N$_2$ was zero in the first cycle and very little N$_2$, which was about 90% lower in its yield obtained with technique-b under identical conditions in the second and third cycles. If liquid N$_2$ is replaced by a dry ice-acetone bath in the second and third cycles, the total yield of N$_2$ is approximately 10% less than what is measured in three freeze-pump-thaw cycles if only dry ice+ acetone bath is used as the refrigerant. Another unusual behaviour was noted in competition studies at dose = 2.16×10^{20} quanta (Table 1) and $[N_2O] \simeq 3.2 \times 10^{-3}M$, the N$_2$ yield is unusually high and does not fit the competition plot. Hence the rate constant ratio was determined from the linear portion of the plot. This effect is not seen under identical conditions except change of total dose to 1.08×10^{20} quanta.

The kinetic analysis of the I$_2$ yields measured from the degassed and non-degassed solutions was attempted using Eq. (3). The plots showed high scatter, however, the accuracy of the measurement of I$_2$ is within $\pm 2\%$. The data from competition studies also showed that neither the degassed nor the non-degassed I$_2$ yield is equal to the corresponding N$_2$ yield. The summation of either the non-degassed or the degassed F$^-$ and N$_2$ quantum yields is not constant and increases with the increase of [scavengers].

Interestingly, deep blue colour is observed when the solution after photolysis in competition experiments is frozen in dry ice-acetone bath. This colour is definitely not due to I$_2$ because synthetic solutions under identical conditions which were not photolysed, did show only the orange brown colour of I$_2$ when frozen in dry ice-acetone bath.

Fe(CN)$_6^{4-}$ system — The photolysis of Fe(CN)$_6^{4-}$ system was carried out using 2537 Å source. The F$^-$ yields obtained from the competition experiments were analysed by the reciprocal plot given in Eq. (2). The best intercept value gave $\Phi(F^-)/6 = 1.05 \pm 0.1$, and $k_1/k_2 = 1.8 \pm 0.15$ which is in good agreement with the value 1.67 obtained by Asmus and Fendler[8]. The value of rate constant ratio is different from that obtained in the I$^-$ system (Table 1). The quantum yield of Fe(CN)$_6^{3-}$ in all these competition experiments was constant having a value 1.84 ± 0.05.

The quantum yields of Fe(CN)$_6^{3-}$ and F$^-$ as a function of [SF$_6$] appear to drop after $1.4 \times 10^{-3}M$ SF$_6$, and the maximum value at this [SF$_6$] of $\Phi[Fe(CN)_6^{3-}]/2 = 1.0 \pm 0.05$, $\Phi(F^-)/6 = 0.68 \pm 0.02$ were calculated. The yield of F$^-$ versus dose plot at $1.4 \times 10^{-3}M$ SF$_6$ versus irradiation time of 4 min is linear and then becomes non-linear (up to 10 min). From the initial linear portion $\Phi(F^-)/6 = 0.68 \pm 0.02$ was obtained. The yield of Fe(CN)$_6^{3-}$ versus dose plot shows a continuous curvature and from the initial slope, $\Phi[Fe(CN)_6^{3-}]/2 = 1.04 \pm 0.005$ was obtained.

KCl, KBr and KCNS systems — These systems were photolysed in presence of $2.4 \times 10^{-4}M$ or $1.4 \times 10^{-3}M$ SF$_6$ using 1849 Å light. In the former case the F$^-$ yields were linear with dose and from the slopes of such plots $\Phi(F^-)/6 = 0.22 \pm 0.005$, 0.17 ± 0.005, and 0.30 ± 0.005 respectively were obtained in KCl, KBr and KCNS systems. In the latter case the F$^-$ yields were linear with dose in the KCNS system but non-linear in KCl and KBr systems, non-linearity being more pronounced and increasing with dose in the KCl system. $\Phi(F^-)/6 = 0.43 \pm 0.02$, 0.33 ± 0.02, from the initial slopes and 0.345 ± 0.005 from the slope of the yield-dose plots were obtained in the KCl, KBr and KCNS systems respectively.

Discussion

Flash photolysis experiments by Matheson *et al.*[17] with Cl$^-$, Br$^-$, I$^-$, CNS$^-$ and Fe(CN)$_6^{4-}$ suggest that with the exception of CNS$^-$, photolysis of these ions leads to the formation of e_{aq}^-. This conclusion is also reached[11,14,17,22] from the known steady state photolysis of Cl$^-$, Br$^-$, I$^-$, and Fe(CN)$_6^{4-}$. In fact, e_{aq}^- concept is well established and widely accepted in the literature because it has theoretical and seemingly experimental justifications known over the past decade or so. The results obtained in the present study do not support fast and diffusion controlled reactions of e_{aq}^- with N_2O and/or SF_6. Assuming the reasonably accepted homogeneous kinetic basis of these reactions, Eqs. (1)-(3) are perfectly valid and hence the same values of k_1/k_2 and $\Phi^\circ(e_{aq})$ must be obtained using these equations:

$$e_{aq}^- + N_2O \xrightarrow{K_2} N_2 + O^- (+H_2O \to OH + OH^-) \qquad \ldots(4)$$

$$e_{aq}^- + SF_6 \xrightarrow{K_1} SF_5^- + F^\circ \qquad \ldots(5)$$

Reactions (4) and (5) are pseudo first order over the studied range of N_2O and SF_6 and hence the rates of these reactions should be a function of SF_6/N_2O only. Thus $\Phi(F^-)/6 + \Phi(N_2)$ should be constant and the summation should be equal to the quantum yield of e_{aq}^- obtained from the intercept. This value should represent the $\Phi^\circ(e_{aq}^-)$ at infinite [scavenger] which is not seen from the present study (Fig. 1). There should be no difference in the degassed from the non-degassed F^- or I_2 yields. All these arguments combined with no formation of N_2 in the first freeze-pump cycle (technique-a) do not support mechanism of e_{aq}^- reactions, and question the existing vital assumptions accepted in arriving at such mechanisms namely: (a) Is the rate of disappearance of e_{aq}^- in the presence of SF_6 and/or N_2O is also the rate to yield the reaction products? (b) Is the species namely e_{aq}^- paramagnetic? (c) Is the band with the absorption maxima $\simeq 7200$ Å due to e_{aq}^- (ref. 23)? Following alternate explanations are proposed in order to interpret the chemical data obtained from the present work: (i) The species e_{aq}^- is not structureless but has definite structure. (ii) This species has structure forming characteristics. (iii) Molecules which form structures with this species can be classed as structure makers and which break the structure of this species can be classed as structure breakers. Radicals are proposed as structure breakers only. (iv) N_2O and SF_6 are both structure makers but SF_6 is proposed as structure breaker relative to N_2O.

Henceforth, the structure with N_2O shall be called as (A) and with SF_6 as (B). The basis of these four hypotheses can be justified from the present work and the information available from the literature. The stable species which may be represented as negative water cluster (NWC) $n(H_2O)^{2-}$, has been proposed earlier by the author[24] and the structures (A) and (B) can be understood as the dipole-dipole interactions between NWC and N_2O, NWC and SF_6 respectively invoking the electrostatic model[24]. It appears that structures (A) and (B) are formed in competition with each other depending on the $[N_2O]$ and $[SF_6]$ and the rate of formation of the structures is different from the rate of their breaking because these are two completely different processes, independent of each other. The post-irradiation breaking of structures is supported from the differing degassed and non-degassed F^- or I_2 yields. The decrease in $\Phi(F^-)$ after about $1.2 \times 10^{-3} M$ $[SF_6]$ (Fig. 2) is interpreted as the rate and mechanism of the breaking of structure B may be different from the low and intermediate $[SF_6]$ or I° acting as structure breaker of NWC which is favoured by SF_6 through collision with I° at high concentrations. The decreasing yield of F^- with the increase in $[SF_6]$ and dose (Fig. 3) supports the view that SF_6 helps I° through collision in breaking the structure

of NWC which is more pronounced as the $[I^\circ]$ increases with dose. The pronounced breaking of the NWC by I° or by its products with I^- also explains the low yield of I_2 in the various KI-SF_6 systems. The increase of I_2 or N_2 yields with increasing $[N_2O]$ (Fig. 1) indicates the nonhomogeneous distribution of the radicals and the rate of the reaction of I° with I^- to form I_2^- may be faster than assumed on the basis of homogeneous kinetics so that the time for this reaction to take place is less than the time for the initial non-homogeneous distribution of I° to attain homogeneous distribution. Thus, the environmental distribution of N_2O shall control the formation of structure (A), i.e. in a region if N_2O is enough to prevent the reaction of I° with NWC (I° acting as structure breaker), the formation of structure (A) shall be favoured.

No formation of N_2 in the first cycle in technique-a is due to the fact that at $-196^\circ C$, N_2O does not bubble out of the solution (freezing point of N_2O is $-120^\circ C$). Perhaps during thawing in the second and third cycles some N_2O bubbles out of the solution due to disturbance of the equilibrium of the system and also some changes in the structure (A) may have been brought about due to freezing with liquid N_2 and thawing with hot water so as to form irreproducible yields of N_2. The yields of N_2 or I_2 dependent upon techniques-b and c and degassed or non-degassed at $1.5 \times 10^{-2} M$ N_2O provides an evidence of the post-irradiation breaking of structure (A). The breaking of structure (A) results in intermediates which react with I^- to change the quantum yield of I_2 to 0.265 ± 0.005 and thus $\Phi(I_2)$ may also be dependent not only on the technique used but also on the $[I^-]$ used for photochemical work.

In the presence of small amounts of O_2, the abnormal increase in F^- yield may be due to O_2 helping SF_6 in breaking the structure of NWC and thus there may be no formation of stable structure (B) in the presence of O_2. The breaking of structures (A) and (B) has been proposed through their intermediates N_2O^- and SF_6^- respectively[25].

The linear increase in F^- yield to time of irradiation = 4 min in the photolysis of $Fe(CN)_6^{4-}$-SF_6 system at $[SF_6] = 1:4 \times 10^{-3} M$ can be interpreted as the formation of structure (B) proportionate to dose and after this time, the rate of structure breaking process may exceed due to increasing amount of dose producing greater concentration of $Fe(CN)_6^{3-}$ which is proposed as structure breaker of NWC and hence a non-linear increase of F^- yield with dose after 4 min. The increasing $[SF_6]$ shall favour the formation of structure (B) until the rate of structure breaking process exceeds that of structure making due to corresponding increase in the concentration of structure breaking products and perhaps to their faster rate of reaction with NWC and hence an apparent decrease of yields is observed after about $[SF_6] = 1.4 \times 10^{-3} M$. The continuous decrease of the $Fe(CN)_6^{3-}$ yield with dose supports the view that $Fe(CN)_6^{3-}$ acts as structure breaker of NWC to yield $Fe(CN)_6^{4-}$ or may also yield monoaquo complex $[Fe(CN)_5.OH_2^{3-}]$ and CN^- ion.

The linearity and non-linearity of F^- yields in the halide and CNS^- systems is explained due to SF_6, depending upon its concentration and the steady state concentration of radicals Br°, Cl° and CNS°, produced on photolysis of their corresponding anions helps the radicals through collision in breaking the structure of NWC. This phenomenon should naturally be dependent upon the nature of the radical also.

Acknowledgement

The author is grateful to the University of Windsor, Windsor, Ontario, for providing the graduate teaching assistantship from 1966-1969, when this work was done in its Chemistry Department. The author also appreciates very much the financial assistance provided to him by his present employer, The Kerr Manufacturing Co., 28200 Wick Road, Romulus, Mich. 48174, which was a great help and encouragement in the presentation of this paper at the 162nd ACS National Meeting in Washington, DC.

References

1. SINGAL, R. K., *Chem. Engng News* (ACS publication; 1 March 1971), 3.
2. (a) SINGAL, R. K., Paper presented in part at the 160th ACS National Meeting in Chicago, USA (Paper No. 130, Physical Chemistry Division).
   (b) SINGAL, R. K., Paper presented in full at the 162nd ACS National Meeting in Washington, D C (Paper No. 48, Division of Food and Agricultural Chemistry).
3. AVERY, E. C., REMKO, J. R. & SMALLER, B., *J. chem. Phys.*, **49** (1968), 951.
4. HEAD, D. A. & WALKER, D. C., *Can. J. Chem.*, **45** (1967), 2051.
5. OHNO, S., *Bull. chem. Soc. Japan.*, **40** (1967), 776.
6. DANIELS, M. & WIGG, E., *J. phys. Cehm. Ithaca*, **73** (1969), 3707.
7. DANIELS, M., *J. phys. Chem. Ithaca*, **73** (1969), 3710.
8. ASMUS, K. D. & FENDLER, J. H., *J. phys. Chem. Ithaca*, **72** (1968), 4285.
9. FRIEDMAN, H. L., *J. Am. chem. Soc.*, **76** (1954), 3294.
10. HATCHARD, C. G. & PARKER, C. A., *Proc. R. Soc.*, **A235** (1956), 518.
11. DAINTON, F. S. & FOWLES, P., *Proc. R. Soc.*, **A287** (1965), 295, 312.
12. OHNO, S., *Bull. chem. Soc. Japan*, **40** (1967), 1770.
13. AWTREY, A. D. & CONNICK, R. F., *J. Am. chem. Soc.*, **73** (1951), 1842.
14. DAINTON, F. S. & LOGAN, S. R., *Proc. R. Soc.*, **A287** (1965), 281.
15. (a) DURST, R. A. & TAYLOR, J. K., *Analyt. Chem.*, **39** (1967), 1483.
    (b) SRINIVASAN, K. & RECHNITZ, G. A., *Analyt. Chem.*, **40** (1968), 509.
16. MEGREGIAN, S., *Analyt. Chem.*, **26** (1954), 1161.
17. JORTNER, J., OTTOLENGHI, M. & STEIN, G., *J. phys. Chem. Ithaca*, **66** (1962), 2029, 2037, 2042.
18. JORTNER, J., OTTOLENGHI, M. & STEIN, G., *J. phys. Chem. Ithaca*, **68** (1964), 247.
19. MATHESON, M. S., MULAC, W. A. & RABANI, J. A., *J. phys. Chem. Ithaca*, **67** (1963), 2613.
20. AIREY, P. L. & DAINTON, F. S., *Proc. R. Soc.*, **A291** (1966) 340, 478.
21. SHIROM, M. & STEIN, G., *Nature Lond.*, **204** (1964), 778.
22. OHNO, S., *Bull. chem. Soc. Japan.*, **40** (1967), 1779.
23. BOAG, J. W. & HART, E. J., *J. Am. chem. Soc.*, **84** (1962), 4090.
24. SINGAL, R. K., *Indian J. Chem.*, **9** (1971), 724.
25. SINGAL, R. K., *Abstracts of papers* (Paper No. 48), Division of Food & Agricultural Chemistry, 162nd ACS National Meeting, Washington DC.

were developed from this basic model using the thermodynamic relations between these properties and the heat capacity.

$$(H° - H°)/T = A + BT + CT^2 + D\ln T/T + E/T + F/T^2$$
$$S° = A + BT + CT^2 + D\ln T + E/T + F/T^2$$

The respective constants of these models are reported in Table 3. Analytical models were not developed for $\Delta H_f°$ and $\Delta G_f°$ because of the discontinuities in the reference thermal functions for I_2 and Br_2 caused by the change in standard reference state to gas from solid and liquid respectively, in the temperature range involved.

These calculated values for BrF_5 and IF_5 may be compared with those reported by Stull[9]. For BrF_5 the agreement between the two sets of values is satisfactory, the largest difference of about 1 cal being in the entropy value. However, for IF_5 the difference is considerably large in $\Delta G_f°$ and log K_f. This is primarily due to the large difference of about 4 kcal in the enthalpy of formation in our calculations and that of Evans et al.[10] used in JANAF tables. For completion of this series, ClF_5 has been included and the enthalpy, Gibbs energy and the equilibrium constant of formation have also been calculated which is an extension of an work of Bougon et al.

The authors thank the Computer Centre, IIT, Kanpur, for making the IBM 7044 available for these calculations and to the National Standard Reference Data Programme of the United States National Bureau of Standards for a research grant under their special International Programme.

References

1. BEGUN, G. M., FLETCHER, W. H. & SMITH, D. F., J. chem. Phys., **42** (1965), 2236.
2. BURKE, T. G. & JONES, E. A., J. chem. Phys., **19** (1951), 1611.
3. STEPHENSON, C. V. & JONES, E. A., J. chem. Phys., **20** (1952), 1830.
4. McDOWELL, R. S. & ASPREY, L. B., J. chem. Phys., **37** (1962), 165.
5. HORAK, J. P., M.S. thesis, Vanderbilt University, 1963.
6. LORD, R. C., LYNCH, M. A. (Jr), SCHUMB, W. C. & SLOWINSKI, E. J. (Jr), J. Am. chem. Soc., **72** (1950), 522.
7. GILLESPIE, R. J. & CLASE, H. J., J. chem. Phys., **47** (1967), 1071.
8. BURBANK, R. D. & BENSEY, F. N., J. chem. Phys., **27** (1957), 982.
9. STULL, D. R. & PROPHET, H., JANAF thermochemical tables (Dow Chemical Co., Midland, Michigan), 1965.
10. EVANS, W. H., MUNSEN, T. B. & WAGMAN, D. D., J. Res. natn. Bur. Stand., **55** (1955), 147.
11. BOUGON, R., CHATELET, J. & PLURIEN, P., C.r. hebd. Seanc. Acad. Sci. Paris, **264** (1967), 1747.
12. ZWOLINSKI, B. J., WILHOIT, R. C., Heats of formation and heats of combustion, American Petroleum Institute Research Project 44 and Thermodynamics Research Centre Data Project (Thermodynamics Research Centre, Texas A & M University, College Station, Texas), 1968.
13. Tech. News Bull. natn. Bur. Stand., **47** (1963), 175.
14. Report of the International Commission on Atomic Weights, J. Am. chem. Soc., **75** (1962), 860.
15. BAUER, H. F. & SHEEHAN, D. F., Inorg. Chem., **6** (1967), 1736.
16. BISBEE, W. R., HAMILTON, J. V., GERHAUSER, J. M. & RUSHWORTH, R., J. chem. Engng Data, **13** (1968), 382.
17. ROGERS, H. H., CONSTANTINE, M. T., QUAGLINO, J. (Jr), DUBB, H. E. & OGIMACHI, N. N., J. chem. Engng Data, **13** (1968), 307.

Radiolysis of KI-N₂O System*

R. K. SINGAL

Polymer Institute, University of Detroit
Detroit, Michigan 48221

Manuscript received 7 December 1970; revised manuscript received 7 April 1971

^{60}Co-γ-radiolysis of aq. KI solution (5×10^{-3}M) in the presence of N_2O has been reinvestigated. Certain discrepancies have been observed in $G(N_2)$ and $G(H_2)$ values as compared to these reported in literature. It is observed that in liquid nitrogen $G(N_2)$ and $G(O_2)$ are zero in the first freeze-pump cycle but not in the second and third cycles. In addition, it is found that $G(N_2)$ and $G(O_2)$ are dependent upon the refrigerant used after γ-radiolysis. This has been suggested to be due to the formation of structures between N_2O and negative water cluster.

GAMMA radiolysis of deaerated millimolar aqueous I⁻ solutions at natural pH has been well studied in the presence of N_2O (ref. 1-4). The system was rechecked for its gaseous and solution products due to some novel features observed by the author in his radiation-chemical study on aqueous systems. In this study, the yields of the gaseous products, e.g. $G(H_2)$, $G(N_2)$, disagree with Buxton and Dainton[2] who used similar experimental techniques and also with Hayon[3], who directly injected the sample for analysis of the gaseous sample from the glass ampoules used for radiolysis. In addition, some odd behaviour of the yields of N_2 and O_2 dependent upon the refrigerant of aqueous solution after gamma radiolysis is also being reported.

Cells, pipettes and storage flasks were cleaned by rinsing them successively with permanganic acid, distilled water, hydrogen peroxide containing nitric acid and finally four times with triply distilled water. Fisher grade KI was used and a solution (5×10^{-3}M, 10 ml) was pipetted into the clean cell and deaerated by three to four freeze-pump-thaw cycles. Desired pressures of N_2O (Matheson grade) were obtained within the cell and their concentration calculated from known solubility data[5]. The system was thermally equilibrated at room temperature before irradiation. Dose rates were measured by the Fricke dosimeter using $G(Fe^{3+}) = 15.6$ ions per 100 eV for ^{60}CO-gamma rays and a dose rate of 7.6×10^{20} eV/litre/min was calculated. After radiolysis, the solution was frozen with liquid nitrogen (technique a), dry ice + acetone (technique b) and no refrigerant (technique c), and the gases were collected into the pre-calibrated capillary of the toepler pump in three cycles. From the capillary, the gases were collected into a bulb which was subsequently hooked on to the mass spectrometer for determining the composition of gaseous products and the limit of reliable reproducibility achieved was ± 1-2%.

No products were formed on radiolysis of deaerated triply distilled water, both at low and high doses;

---

*The work reported herein was carried out at the Atomic Energy Ltd, Chalk River, Ontario, Canada.

TABLE 1 — RADIOLYSIS OF KI-N_2O SYSTEM

{[N_2O] = $2.0 \times 10^{-2}M$]}

Product	Technique a	Technique b	Technique c
$G(H_2)$	0.66 ± 0.02	0.65 ± 0.02	0.65 ± 0.02
$G(N_2)$	0.99 (No N_2 in first cycle)	4.1 ± 0.1	3.2 ± 0.1
$G(O_2)$	0.09 (No O_2 in first cycle)	0.63 ± 0.01	0.33 ± 0.01

'flickering clusters' model for the structure (I) of H_2O

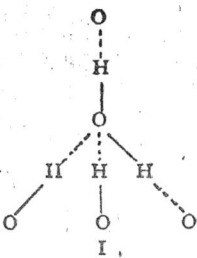

Due to the large difference in electronegativity of H and O, the charge shall flow towards the more electronegative O which shall leave the hydrogens in the water cluster charge deficient and the extra charge produced on gamma radiolysis of aqueous solutions can be localized on these charge deficient hydrogens. Considering a single H_2O molecule in the cluster, the localization of charge on both hydrogens shall make the cluster diamagnetic and the species can be described as [$n(H_2O)^{2-}$]. Only dipole-dipole interactions of N_2O with the negative water cluster are proposed. Thus, the band having absorption maxima ~7200 Å obtained by Boag and Hart[8] may be interpreted due to the negative water, cluster and visualizing its stability with N_2O, optical transition responsible for this band can be $1s \rightarrow np$. The author has collected enormous experimental evidence against $\bar{e}_{aq}$. mechanisms the details of which will be published elsewhere. Non-homogeneously distributed H and OH formed on radiolysis of aqueous solution are proposed to act as structure breakers of negative water cluster and in a particular region of the aqueous medium; the rate of structure making of N_2O with negative water cluster may exceed that of structure breaking by H and OH so that reactions,

$$H + H \rightarrow H_2 \quad \text{...(1)}$$
$$OH + I^- \rightarrow I^\circ + OH^- \quad \text{...(2)}$$

are favoured in the presence of N_2O over its absence, thus accounting for the increased $G(H_2)$ in the presence of N_2O. $G(I_2) = 1.7 \pm 0.05$ obtained in the present study in the presence of $2 \times 10^{-2}M$ [N_2O] is in good agreement with Anbar et al.[4]. Radiolysis of millimolar I^-_{aq} alone resulted in $G(I_2) = 0.08 \pm 0.005$ and reactions (2-4) may account for the increased $G(I_2)$ in presence of N_2O:

$$I^- + I^\circ \rightarrow I^-_2 \quad \text{...(3)}$$
$$I^-_2 + I^-_2 \rightarrow I^-_3 + I^- \quad \text{...(4)}$$

Structure of N_2O with the negative water cluster is proposed to be stable and breaks only when N_2O bubbles out of the solution due to disturbance of equilibrium of the system either due to freeze-thaw processes of the solution in the reaction vessel or pumping off dissolved or gasphase N_2O. At $-196°C$, N_2O freezes and thus the structure may be stable at liq. N_2 temperature but at $-80°C$ (technique b), N_2O is pumped off (f. pt. of $N_2O = -120°C$). The structure may break through the intermediate N_2O^- yielding N_2 and O_2 in the post-irradiation period due to some complex mechanism. Anbar et al.[4] have suggested the formation of the intermediate N_2OH having a lifetime long enough to permit chemical

time of irradiation 0.5 and 10 min respectively. However, radiolysis of deaerated $5 \times 10^{-3}M$ KI solution gave $G(H_2) = 0.45_5$ which is well accepted and is in agreement with literature value[6].

Zero yields of N_2 and O_2 in the first freeze-pump cycle using technique (a) was always reproducible but not the second and third cycles. However, if liq. N_2 is replaced by dry ice + acetone-bath in the fourth and fifth cycles, excess N_2 and O_2 are obtained and total yield of N_2 is approximately 10% less than if only technique b is used for measurement of the gases. As is evident from the data in Table 1 (technique b), $G(N_2)$ is about 30% more than reported by others[2,3] and $G(H_2)$ approximately 50% higher than obtained by Buxton and Dainton[2]. However, $G(N_2)$ is in good agreement with them if technique c is used.

If the well accepted basis of fast and diffusion-controlled reaction of $\bar{e}_{aq}$ with N_2O is true, the yields of N_2 should be identical using three techniques. Freezing the solution with liq. N_2 before measurement of the gaseous products should not result in any inconsistencies in their G values with other techniques if the gases are already present in solution in equilibrium with the gas phase in the reaction vessel as seen from $G(H_2)$ and not from $G(N_2)$ and $G(O_2)$ (Table 1). It was noted that there is absolutely no problem in the deaeration of aqueous solutions before radiolysis if liq. N_2 is used as the refrigerant, rather it is seen that this method is more efficient for outgassing of dissolved N_2 and O_2 as compared to techniques b and c.

Because $G(H_2) = 0.45_5$ obtained on radiolysis of deaerated millimolar KI solution is in agreement with the literature value and all experimental details of measurement and analysis of gaseous products were carefully checked and rechecked, no experimental error is comprehended in the gaseous products. Thus, it is very likely that the dependence of N_2 and O_2 yields on the techniques used for their measurement is due to chemical reasons. This claim may be supported by the spectrum of inconsistent results reported in the past decade or so in the literature[7].

It is proposed that N_2 and O_2 may be produced by the post-irradiation breaking of structures formed by N_2O with $\bar{e}_{aq}$ (ref. 8). The species proposed is negative water cluster which should be diamagnetic in nature. Frank and Wen[9,10] have proposed a reaction. The formation of intermediates like N_2O^- or N_2OH in aqueous radiation chemistry is subject to great speculation and has no basis other than the assumption that they could not be detected by fast reaction techniques due to their short lifetime. The proposal of formation of structures may provide an answer to all the unsuccessful attempts at solving the question of intermediates by fast reaction techniques.

The proposal of structures of N_2O with negative water cluster when confirmed spectroscopically should open a new era in structural chemistry.

The author expresses his appreciation for the help and encouragement rendered by Dr W. A. Seddon of the Atomic Energy of Canada Ltd, Chalk River, Ontario, where this work was done in the summer of 1968.

References

1. SINGAL, R. K., Ph.D. thesis, University of Windsor, Windsor, Ontario, Canada, 1970, paper presented at the 160th ACS National Meeting.
2. BUXTON, G. V., DAINTON, F. S., Proc. R. Soc., A287 (1965), 427.
3. HAYON, E., Trans. Faraday Soc., 61 (1965), 723.
4. ANBAR, M., MEYERSTEIN, D. & NETA, P., J. phys. Chem., 68 (1964), 2967.
5. Solubilities of inorganic and metal-organic compounds, Vol. II, edited by W. F. Linke (American Chemical Society, Washington), 1965, 794.
6. ALLEN, A. O., The radiation chemistry of water and aqueous solutions (D. Van Nostrand Co. Inc., New York), 1961.
7. CZAPSKI, G., Advances in chemistry Series 81; Radiation chemistry, Vol. 1 (American Chemical Society, Washington), 1968, 106-28.
8. BOAG, J. W. & HART, E. J., J. Am. chem. Soc., 84 (1962), 4090.
9. FRANK, H. S. & WEN, W. Y., Disc. Faraday Soc., 24 (1957), 133.
10. FRANK, H. S., Proc. R. Soc., A247 (1958), 481.

A Kinetic Study of the Oxidation of Steroid Alcohols by N-Bromosuccinimide

N. VENKATASUBRAMANIAN & N. S. SRINIVASAN

Department of Chemistry, Vivekananda College, Madras 4

Manuscript received 15 March 1971

The rates of NBS oxidation of epimeric pairs of alcohols, viz. cholestan-3α- and -3β-ols and of 3α- and 3β-hydroxy-5β-pregnane-20-ones, show little difference, indicating that there is very little selectivity in the NBS oxidation of conformational epimers.

THE high selectivity of N-bromosuccinimide in the oxidation of steroid alcohols is a well-documented phenomenon[1-3]. However, very few kinetic measurements of this reaction have been carried out. Although Langbein and Steinert[4] while studying oxidation of 3,17-α-dihydroxy-21-bromopregnane-11,20-dione by NBS concluded that the oxidation involved a cyclic transition state, embodying a rate-determining loss of the hydrogen at the secondary carbon as a hydride ion, it must be pointed out that only one conformer was used for this reaction. Also these investigators followed only the bromine oxidation and not the NBS oxidation as their

TABLE 1 — SECOND ORDER RATE CONSTANTS FOR THE NBS OXIDATION OF CHOLESTAN-3α- AND -3β-OLS

{Solvent (%, vol./vol.): tert-Bu-OH, 60% + HOAc, 25% + H_2O, 15%; [alcohol] = 0·0015M; [NBS] = 0·0008M; [HgOAc$_2$] = 0·001M}

Alcohol	$k_2 \times 10^1$ (litre mole^{-1} sec^{-1})		
	30°	40°	45°
Cholestan-3α-ol	5·63	8·25	11·6
Cholestan-3β-ol	4·72	6·90	9·51

TABLE 2 — SECOND ORDER RATE CONSTANTS FOR THE NBS OXIDATION OF SOME STEROID ALCOHOLS

{[Solvent (%, vol./vol.): HOAc, 70% + H_2O, 30%; [alcohol] = 0·0016M; [NBS] = 0·00084M; [HgOAc$_2$] = 0·001M}

Alcohol	$k_2 \times 10^2$ (litre mole^{-1} sec^{-1})		
	45°	50°	60°
3α-Hydroxy-5β-pregnane-20-one	1·52	2·44	4·17
3β-Hydroxy-5β-pregnane-20-one	2·18	3·57	6·21
3α-Hydroxy-5α-androstane-17-one	1·11	2·16	3·00

reactions were carried out in the presence of an initially added catalytic quantity of Br_2 (the Br$^-$ produced subsequently reacted with NBS to give more Br_2).

It has been demonstrated that pure NBS oxidations (free from concurrent Br_2 oxidations) can be carried out in the presence of added mercuric acetate and that the oxidation proceeds by a cyclic mechanism involving the alcohol and the NBS molecule[5,6]. We present, in this communication, the results of the NBS oxidation of two stereo-isomeric pairs of steroid alcohols (cholestan-3α- and 3β-ols and 3α- and 3β-hydroxy, 5β-pregnane-20-ones) in acetic acid-water and acetic acid-tert-butanol-water mixtures as solvents. The reactions were followed by the iodometric assay of unreacted NBS at various time intervals[5]. The pseudo unimolecular rate constants for these reactions are collected in Tables 1 and 2. The data for the oxidation of 3α-hydroxy-5α-androstan-17-one are also included for purposes of comparison.

It is observed that in the case of both the epimeric pairs of alcohols there is very little difference in the rates of oxidation although in both cases the isomer with an axial hydroxyl (cholestan-3α-ol and 3β'-hydroxy-5β-pregnane-20-one) is oxidized slightly faster than its conformer. It is also interesting to note that even the 3-α-hydroxy-5α-androstan-17-one with the more stable A-B trans fusion does not differ much in reactivity from 3α-hydroxy-5β-pregnane-20-one. It has generally been observed that reactions that involve a rate limiting attack on a group equatorially or axially disposed usually show much greater differences in reactivity. For example, in the chromic acid oxidation of the isomeric cholestan-3-ols, the 3α-isomer is oxidized about 3 times faster than the 3α-isomer[7].

Although there is a large selectivity in the NBS oxidation of compounds depending on the position of the groups attacked, it seems that there is little

U.S. and the first use of this new method will be to stockpile rare blood types and to meet emergency situations.

G. A. Jamieson, Ph.D.
Research Director, Blood Program, American National Red Cross, Bethesda, Md.

Porfiromycin patent

SIRS: In C&EN for Nov. 23, 1970, page 21, you refer to the recent ruling by the House of Lords in American Cyanamid Co. *vs.* Upjohn Co.

I submit that your review of this decision leaves a false impression. The microorganism involved in this litigation, *Streptomyces verticillatus*, was always identified. In fact, Lord Reid stated: "The description in the specification is not attacked as being in any way insufficient or misleading."

The issue involved in this litigation was the availability of the microorganism to others. The ruling of the House of Lords was that American Cyanamid did not have to make available to others cultures of the microorganism.

John V. Whittenburg
Patent Counsel, American Cyanamid Co., Stamford, Conn.

Inspires students

SIRS: Thanks to C&EN for including the article "Albert Ghiorso—Element Builder" by John F. Henahan (Jan. 18, page 26). The conversational style depicted the humanness of this scientist, a quality which I try to emphasize when teaching junior college chemistry for the nonscience major. Also, the search for transuranium elements was summarized and presented clearly, making this article a good one to be recommended for student perusal.

This article and others on topics including nuclear chemistry, pollution, drugs of abuse, chemical research, and food additives provide good reference materials for students in my introductory chemistry class. Students in small discussion groups include C&EN articles as reference sources for background reading and then discuss pros and cons of these timely topics. As a result, student interest in the course has increased considerably.

Ralph A. Burns
Chemistry Instructor, East Central Junior College, Union, Mo.

Negative water cluster

SIRS: The proposal of "Negative Water Cluster" (C&EN, Nov. 9, 1970, page 8) is novel and unique. It challenges the concept of "hydrated electron," which has been well studied in radiation chemistry in the past decade or so and is widely accepted in the chemical literature. The idea of "negative water cluster" [n(H$_2$O)$^{2-}$] has been described in my Ph.D. thesis submitted on Dec. 9, 1969, to the University of Windsor, Windsor, Ont., and subsequently presented in the 160th ACS National Meeting in Chicago (Division of Physical Chemistry, paper No. 130). For the convenience of scientists who are perhaps not aware of this novelty, I am describing it here in brief and this may be helpful for comments and discussions on this idea.

It appeared impossible to explain the results obtained in the photolysis and gamma radiolysis of aqueous I$^-$ solutions according to the well-accepted basis of fast and diffusion-controlled reactions of hydrated electron ($\bar{e}_{aq}$), and hence I questioned the $\bar{e}_{aq}$ concept and assumed four hypotheses (see abstract) which appear adequate in explaining the results. However, this meant a complete and violent opposition to the fundamentals accepted in radiation chemistry and also serious implications on many sides of structural chemistry. I have proposed a stable species in deaerated water instead of a fast-reacting $\bar{e}_{aq}$, produced on photolysis or radiolysis of I$^-_{aq}$. A "flickering clusters" model has been proposed for the structure of water [*Proc. Roy. Soc., London*, A, 247, 481 (1958)]:

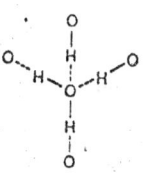

Due to large difference in electronegativity between H and O, the charge shall flow toward the more electronegative O, which shall leave the hydrogens in the neutral water cluster charge deficient and hence the extra charge produced on photolysis or radiolysis of aqueous solutions can be localized on these charge-deficient hydrogens. Considering a single water molecule in the cluster, the localization of charge on both hydrogens shall make the cluster diamagnetic. SF$_6$ or N$_2$O forms stable structures with the negative water cluster which have been proposed by me to break in the postirradiation period through their intermediates SF$_6^-$ and/or N$_2$O$^-$.

What do I mean by further investigations of negative water cluster and its possible correlation with anomalous water? It is well known in radiation chemistry that alkali and alkaline earth metals lose charge to water, e.g.,

$$M^\circ + H_2O \rightarrow M^+ + \bar{e}_{aq}$$

and radiation chemists have been tracking down this so-called short-lived species by all sorts of sophisticated optics. My hope is that if the idea of negative water cluster is reasonable, then one may be able to observe this species by infrared and/or Raman spectroscopic techniques after dissolving alkali and alkaline earth metals in deaerated water. This line of experimentation seems reasonable because of the nature of ions reported as impurities in anomalous water, believed by some to come from the leaching of glass surface. If the spectral bands are found close to what have been reported in recent years in anomalous water, then the chemist may be able to give a start to the most astounding controversy in the chemical literature.

R. K. Singal
Polymer Institute, University of Detroit

Priority for eco-problems

SIRS: Mr. Ehntholt's letter (C&EN, Jan. 11, page 6) in response to Dr. Wolman (C&EN, Nov. 30, 1970, page 5) states that chemists "should have been performing ... research on ... what would happen to a lake if we discharged small amounts

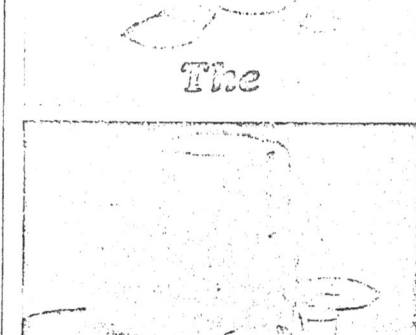
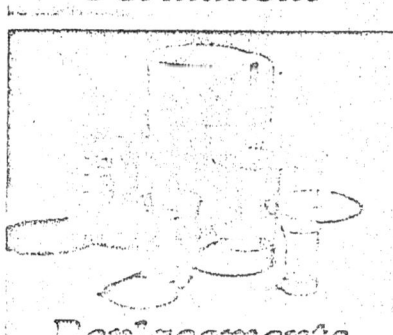

man or woman on the street that no scientist can really be trusted to carry out his or her work safely and honestly if either glory or profit looms large on the horizon? Such a message does nothing to improve the badly tarnished image of chemistry.

Woodfin V. Ligon Jr.
Schenectady, N.Y.

SIR: I want to put my two cents' worth of ideas into the recent controversy regarding the cold fusion process reported by B. Stanley Pons and Martin Fleischmann (C&EN, April 10, page 6). It has been reported in the local press that both chemists had to spend $100,000 of their own money in the past four or five years in order to accomplish what they have reported. They had to choose this route because they knew when they found this unusual cold fusion process years ago that they would face all kinds of hurdles and skepticism from the experts in their field to have their novel ideas approved and funded for further research. They are still facing skepticism and disbelief from the experts in the field. As I have read in the press accounts, for cold fusion to be possible, as claimed by Pons and Fleischmann, one will have to rewrite the laws of physics and chemistry.

Problems with the thinking process of the experts are that they continue to practice the ideology that "experiment guides and theory decides," so they immediately raise hell over some novel experimental findings that the theory cannot explain. Their attitude must be condemned, and these two chemists must be congratulated on their brilliant discovery whether it can be explained by cold fusion or not. Skepticism about novel experimental findings is a natural human reaction that creates more discussion about such experimental truth that ultimately results in a discovery beneficial to mankind. But to destroy such findings with their strongly held theoretical views must be discouraged.

I predict from the published literature of radiation chemistry and from unpublished work done by me during my Ph.D. research about 20 years ago that processes of cold fusion can be better controlled in the absence of environmental O_2 in the electrolysis of D_2O in the presence of palladium and platinum as electrodes. The lifetime of D_2O at the negatively charged palladium core can be enhanced in the absence of competing oxygen molecules for negative charges. I do not have access to the research papers of Pons and Fleischmann,

but water and deuterium oxide must be thoroughly understood—which I claim that they are not according to present theories we know in physics and chemistry. I hope that someday *The Journal of Physical Chemistry* and/or the *Journal of the American Chemical Society* comes around to publishing my research which was communicated to these journals by me about 20 years ago. It can generate a whole new excitement about fusion processes in nature and may be able to better explain the cold fusion claims of Pons and Fleischmann. Perhaps Congress can write some laws about the negative attitudes of experts in the field of science so that new ideas are not pushed behind and so that no scientist has to spend his or her own hard earned personal money to prove claims to the scientific world. I hope that the American Chemical Society takes up this matter soon, and writes very strict guidelines in the publication of novel research no matter where it comes from. I can only hope that experts in their chosen field of science start practicing "theory guides and experiment decides" and not the other way around.

Rajendar K. Singal
Las Vegas, Nev.

SIR: According to your News of the Week segment on nuclear fusion (C&EN, April 3, page 5), the "University of Utah in its news release paints a rosy picture of the electrolytic fusion process as a relatively simple source of abundant, clean energy."

The electrolytic fusion process described is not an abundant source of energy. Each year, 70 to 80 tons of palladium (Pd) reach free-world markets (*National Geographic*, Nov. 1983, page 692). Eighty tons equals 72,574,800 g.

The density of Pd is 12.16 g per cc. Pons told you that 26 watts per cc of palladium has been obtained. This means that 0.4676923 g of Pd produces 1 watt, and 80 tons will produce 155.176 MW. This is enough for only 117,114 people (energy-people relationship derived from *National Geographic*, March 1982, page 408).

On page 5 of C&EN, April 3, you state: "Six days after the university's announcement, state officials pledged $5 million to expand fusion experiments."

Do the state officials know that Pd is one of the scarce elements? Did they estimate the maximum possible yearly power output? Or did they blindly believe the University of Utah's statement of abundant energy? Maybe they don't care what sort of return on investment taxpayers get.

After decades of work, plasma physicists have spent billions of taxpayer dollars (about one third of a billion dollars in 1988 to magnetic fusion, over one third of a billion dollars projected for 1989 and 1990—see C&EN, Feb. 20, page 19) building huge plasma containment vessels, magnets, power supplies, and lasers. Like a black hole sucking in gold bars, fusion research has not returned a single cent. Physicists working on fusion had the wrong approach all along.

Maybe a nationwide poll should be taken, asking taxpayers whether or not they want to continue paying one third of a billion dollars a year on magnetic fusion research, or whether they would like all the fusion equipment auctioned off or recycled, with the proceeds of the auctions going back to the taxpayer.

Rolf Johansson
Lewiston, N.Y.

SIR: The recent announcements by two electrochemists, Pons and Fleischmann, that they had performed experiments in an electrolytic cell that causes deuterium atoms to fuse, has attracted a lot of interest and speculation. C&EN ran two articles and the *Wall Street Journal* hardly misses a day without publishing one on deuterium fusion. In general, the reader gets the impression that the new announcement of room-temperature fusion of deuterium was a first, but perhaps it just isn't so. The speculation is very interesting.

Perhaps you would like to call it to your readers' attention that M. L. Olephant, P. Harteuk, and Lord Rutherford published in a 1934 issue of *Nature* that they observed deuterium fusion at room temperature. They found if an ammonium salt was bombarded either with protons or deuterons the only result was that some of the protons were driven out of the salt. But if an ammonium salt containing deuterium was bombarded by deuterons, even at energies as low as 100,000 volts, an enormous evolution of swift protons occurred—more than in any other known reaction—and it was obvious that energy was being liberated in some nuclear change. P. I. Dee [*Nature*, **133**, 564 (1934); *Proc. Roy. Soc.*, **148**, 623 (1935); and with C. W. Gilbert, *Proc. Roy. Soc.*, **149**, 200 (1935)] examined these reactions further and found that the process was very complex, giving (1) protons of 14 cm range (3×10^6 eV; 30 times the energy of the bombarded particles), (2) neutrons, (3) single- and (4) double-charged particles of shorter range; numbers three and four were found to be ^3H and ^3He.

Continued on page 45